Frederick Coombs

Coombs' Popular Phrenology

Exhibiting the Phrenological Admeasurements of above fifty Distinguished

and Extraordinary Personages of both Sexes

Frederick Coombs

Coombs' Popular Phrenology
Exhibiting the Phrenological Admeasurements of above fifty Distinguished and Extraordinary Personages of both Sexes

ISBN/EAN: 9783337306496

Printed in Europe, USA, Canada, Australia, Japan

Cover: Foto ©berggeist007 / pixelio.de

More available books at **www.hansebooks.com**

POPUL[AR]

PHREN[OLOGY]

FIFTY DISTING[UISHED]

O[F]

VARIOUS [

𝕰mbellish[ed

BEAUTY, [

WITH OBSERV[ATIONS

FOWLER

Boston:
142 Washington S[t.

Phrenological Character of Gen. Washington.

The Patriot, Hero, Statesman, here behold,
Whose soul undaunted, and lion courage bold,
Rescued his bleeding country from foreign tyrants' yoke,
And then, (surpassing greatness,) laid his high honors at your feet

Physiognomical Character.

The firm the pure the kind, expressive face,
(Fair mirror of a fairer soul!)
Proc aims to all the world that all is peace within.
Rest! warrior, rest! thy glorious deeds and memory lives
With grateful millions, till time shall be no more.

PREFACE.

In presenting this little work to the public, the author has been impelled by a desire of furnishing some tangible and positive evidence of the truth of Phrenology, and the practicability of a mathematical demonstration of this beautiful science, of practical utility.

It has long been a desideratum to obtain some uniform and certain mode of ascertaining the exact amount of cerebral matter or brain in the several compartments of the skull, invariably assuming the point between the ears as the basis of the superstructure. This the author hopes he has successfully accomplished.

Amongst the tables of admeasurements, are those of nearly fifty remarkable or distinguished persons, contrasted with those of idiots, murderers, &c. There are also nearly as many remarkable and well-authenticated skulls of the various nations of the world.

He begs to express his obligations to those gentlemen who have so kindly permitted their names to be used for the advancement of science. To the ladies, also, he is much indebted; but, with that native modesty which is so characteristic of the American ladies, he could not obtain permission to give their names in full.

Remarkable as many of the skulls may appear, they have all been accurately drawn from well-authenticated natural specimens.

Application of the Phrenometer.

The admeasurements of skulls here presented, it ought to be understood, have not been selected for the purpose of maintaining a favorite theory, but embrace the greater part of F. Coombs's private collection, and may at any time be seen at his office, the remainder being, with one or two exceptions in the Museum of the Boston Medical Society, who will not be accused of any partiality for the science. The author has to express his obligations to the President, Dr. Warren, for his politeness in the use of them. In these tables there is one remarkable feature, which cannot fail to strike the superficial observer. In comparing the intellectual faculties of the Africans, Asiatics, Indians, and Malay variety of the human family, it will be seen, although some were gigantic men, there is not one of them exceeding 4¾ in the organs of comparison and causality; whilst none of the heads of even very diminutive men, but who have distinguished themselves in letters, are so deficient in the above organs. 4½ to 4¾ appears to be a line of demarkation distinctly drawn betwixt superior and inferior heads. The establishment of this fact is probably one of the most important of the present age, and constitutes a new era in the science of Phrenology. It is hoped the results here submitted will stimulate others in the profession to make similar collections, and, if possible, by the same instrument; for which purpose the drawing is given. No phrenologist ough to think of estimating the size or volume of brain, without frequently resorting to this instrument, as it will be found to be of the very greatest importance in forming a correct estimate of the proportions of the head, or phrenological organs. In offering these remarks, the author conceives it to be a duty he owes the public to caution them not to condemn the science from an imperfect examination of the head by mere tyros in the profession. To this instrument he always resorts in doubtful or contested cases. The results obtained give the exact length of fibre from the base of the brain, or point exactly midway between the ears, to the external development of every organ. The temperaments of the individuals are not designated, but of all the distinguished heads he has examined, the whole of them, or nearly so, are strongly marked with the bilious, and rarely with lymphatic. It will be seen that the heads of giants are not corresponding with their extraordinary statures, the Kentucky giant being small as a dwarf in some of his animal organs. The head is somewhat in proportion to

the body, yet it by no means follows that a giant has a gigantic head.* Strictly speaking, these admeasurements give the size of one hemisphere of the brain only, the other, of course, being a complete counterpart, the action in both being simultaneous. In comparing the heads of living persons with the skulls, as a general rule about one eighth of an inch must be allowed to the skulls, to allow for the thickness of the hair, integuments, &c., excepting in the organs of tune, constructiveness, and acquisitiveness, and also amativeness, in which organs the thickness of the muscles cannot be ascertained. We cannot, therefore, in these organs, compare the skulls with the living heads; every other organ may, allowing only the difference as before stated. Language, and some of the perceptive organs, which cannot be measured, are omitted in this scale.

* There is generally an analogous formation betwixt the head and body, in so far as this: where he has observed a head very large in the animal organs, it is usually accompanied with a strong, vigorous, muscular development of body, as admirably illustrated in the forms of the different sexes. It will be almost invariably found, females are as deficient of the combative propensities as of the physical powers necessary to carry them into execution, of which fact persons may easily satisfy themselves by comparing the widths of heads of the opposite sexes, man being usually much the widest in the base around the ears where these organs are located. The admeasurements of heads are also strikingly illustrative of this fact, being almost invariably found wider from ear to ear in destructiveness in male heads of every description, as illustrated by the outlines below. To those desirous of knowing their own exact cerebral organization, these tables are given for the purpose of enabling them to compare themselves critically and accurately with the great variety of remarkable heads given, also as a means of judging of the science.

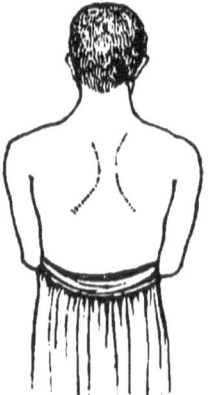 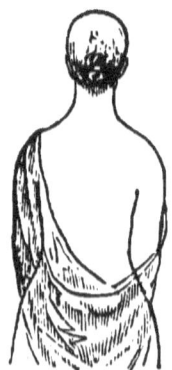

A certain width of head is necessary to give impetus, resolution, and determination of character, in either ladies or gentlemen; and when too wide in either, implies passion and severity. The science in this respect is of incalculable value, enabling us by its assistance to form happy matrimonial alliances, &c. [See page 117.]

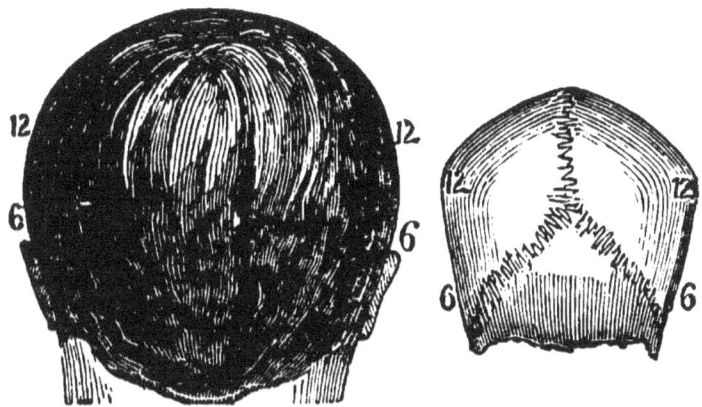

Explanation of Lines, &c. on next page.

No. 1 is the correct drawing of the back-head of a reckless murderer. No. 2 that of a cowardly Hindoo. The lines radiating from the base give the length of the organ 6, or destructiveness, the organ 12, or caution; the perpendicular line of firmness, &c. So in the tables of measures, they extend to every organ except language and the perceptive faculties. By turning the instrument on its axis, the balls resting in the ears, [see the drawing,] we obtain with equal facility the front, back, and side organs, or rather one perfect hemisphere of the brain, to the fractional parts of inches, and this is probably the only plan yet devised for obtaining the length or size of each organ from this centre or base of the brain, admitted by all to be the point from which it may be said all the organs originate.

These tables have been prepared with the most scrupulous regard to accuracy, and have been invariably taken by himself, (except his own head,) with the same beautiful instrument. They are intended for a twofold purpose—first, as a means of accurate comparison between highly civilized, semi-barbarous and savage nations. Secondly, as a standard of comparison for those who may wish to submit themselves to the same rigid test of mathematical demonstration—for which purpose a vacant column is left for such comparison with F. Coombs' approved bust, or perfect head, also with the head of the President of the United States, members of Congress, &c.

Whilst to many these tables may appear somewhat tedious and uninteresting, yet the author conceives, to those who are desirous of investigating the merits of the science, they will prove highly attractive; the exact phrenological character of every skull and person being accurately ascertained by this mode, and every portion of the head presented to view with much greater accuracy than the most costly engravings or descriptions could do.

This mode of proceeding will also tend to disabuse the public mind of the erroneous ideas entertained as to the minute bumps and depressions supposed to be mainly indicative of character, but which rarely exist to any extent; whilst by these tables it will be seen that many heads differ several inches in the same organs. It is this which enables the phrenologist to give such striking evidences of the truth of the science, and not from bumps which it would require a microscopic power to discover.

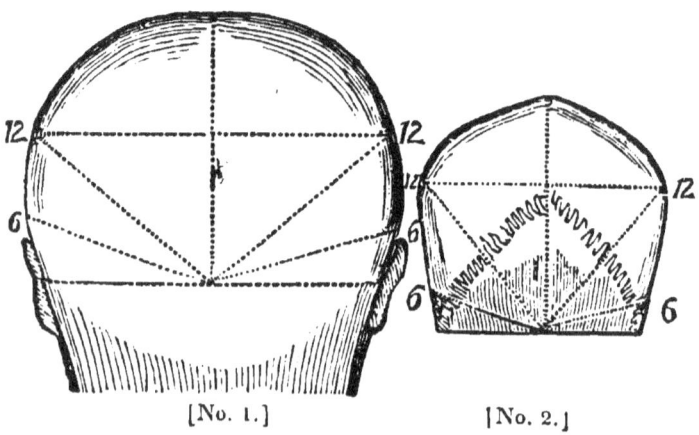

[No. 1.]   [No. 2.]

Admeasurements of Heads of Distinguished Persons.	M. Van Buren, President.	New England's Fairest Flower.	R. M. Johnson, Vice President.	Beautiful Bostonian.	Hon. J.Q. Adams.	The Amiable New Yorker.	Hon. Mr. Preston.	Hon. Mrs. P.	Hon. F. Granger.	The Fair Philadelphian.
1. AMAT	$2\frac{7}{8}$	$2\frac{5}{8}$	$3\frac{3}{8}$	$2\frac{1}{4}$	$3\frac{1}{8}$	$2\frac{3}{8}$	$3\frac{1}{4}$	$2\frac{1}{2}$	3	$2\frac{3}{8}$
2. PHILO.	$4\frac{1}{8}$	$3\frac{7}{8}$	$4\frac{1}{4}$	4	$4\frac{3}{8}$	$3\frac{3}{4}$	$4\frac{1}{4}$	4	$4\frac{3}{8}$	$3\frac{3}{8}$
3. CONCE.	$4\frac{7}{8}$	$4\frac{5}{8}$		$4\frac{3}{4}$	$4\frac{1}{4}$	$4\frac{1}{8}$	$4\frac{5}{8}$	$4\frac{3}{4}$	$5\frac{1}{8}$	$4\frac{1}{8}$
4. ADHES.		$3\frac{7}{8}$	4	$4\frac{1}{4}$	$4\frac{3}{8}$	$3\frac{3}{8}$	$4\frac{1}{8}$	$4\frac{1}{8}$	$4\frac{3}{8}$	$3\frac{3}{8}$
5. COMBA.	3	3	$3\frac{3}{8}$	$2\frac{3}{4}$	$3\frac{3}{8}$	$3\frac{1}{8}$	$3\frac{1}{4}$	$2\frac{7}{8}$	$3\frac{3}{8}$	$2\frac{1}{4}$
6. DESTR.	3	$2\frac{7}{8}$	$3\frac{1}{4}$	3	$3\frac{3}{8}$	$2\frac{5}{8}$	$3\frac{1}{2}$	3	$3\frac{1}{4}$	$2\frac{5}{8}$
ALIM.				$2\frac{13}{16}$						
7. SECRE.	$3\frac{5}{8}$	$3\frac{5}{8}$	$3\frac{1}{2}$	$3\frac{3}{8}$	4	$3\frac{5}{8}$	$4\frac{1}{8}$	$3\frac{1}{2}$	$3\frac{3}{4}$	$3\frac{5}{8}$
8. ACQUI.	4	$3\frac{5}{8}$	$3\frac{1}{4}$	$3\frac{5}{8}$	$4\frac{1}{8}$	$3\frac{3}{4}$	$4\frac{3}{8}$	$3\frac{7}{8}$	$3\frac{3}{8}$	$3\frac{7}{8}$
9. CONST.	$4\frac{1}{4}$	4	4	$3\frac{9}{16}$	$4\frac{1}{8}$	$3\frac{3}{4}$	$4\frac{1}{2}$	$3\frac{3}{4}$	$3\frac{5}{8}$	$3\frac{9}{16}$
10. SELF-E.	$5\frac{3}{8}$	$5\frac{1}{4}$	$5\frac{1}{8}$	$5\frac{1}{4}$	$5\frac{5}{8}$	5	$5\frac{3}{8}$	$5\frac{1}{4}$	$5\frac{5}{8}$	$5\frac{1}{8}$
11. APPRO.	$5\frac{1}{4}$	$5\frac{3}{8}$	$5\frac{1}{8}$	$5\frac{3}{16}$	$5\frac{1}{2}$	5	$5\frac{3}{8}$	$5\frac{1}{8}$	$5\frac{1}{2}$	$5\frac{1}{8}$
12. CAU.	5	$4\frac{5}{8}$	$4\frac{3}{4}$	$4\frac{7}{8}$	5	$4\frac{1}{2}$	$5\frac{1}{2}$	5	$4\frac{7}{8}$	$4\frac{3}{4}$
13. BENEV.	$5\frac{5}{8}$	$5\frac{1}{8}$	$5\frac{5}{8}$	$5\frac{1}{2}$	$5\frac{5}{8}$	$5\frac{1}{4}$	$5\frac{7}{8}$	$5\frac{3}{8}$	$5\frac{7}{8}$	$5\frac{3}{8}$
14. VENER.	$5\frac{5}{8}$	$5\frac{5}{16}$	$5\frac{5}{8}$	$5\frac{7}{16}$	$5\frac{3}{4}$	$5\frac{5}{16}$	$5\frac{7}{8}$	$5\frac{3}{8}$	$5\frac{5}{8}$	$5\frac{5}{16}$
15. FIRM.	$5\frac{3}{4}$	$5\frac{9}{16}$	$5\frac{5}{8}$		$5\frac{7}{8}$	$5\frac{1}{4}$	6	$5\frac{3}{8}$	$5\frac{7}{8}$	$5\frac{1}{4}$
16. CONSCI.	$5\frac{3}{8}$	$5\frac{1}{2}$	$5\frac{5}{16}$	$5\frac{3}{8}$	$5\frac{13}{16}$	$5\frac{3}{16}$	$5\frac{7}{8}$	$5\frac{5}{16}$	$5\frac{3}{4}$	$5\frac{1}{8}$
17. HOPE		$5\frac{3}{8}$				$5\frac{3}{16}$			$5\frac{5}{8}$	$5\frac{1}{4}$
18. MARV.		5								
19. IDEAL.	$4\frac{3}{4}$	$4\frac{7}{8}$	5	$4\frac{3}{4}$	$5\frac{1}{4}$	$4\frac{1}{2}$	$5\frac{1}{2}$	$4\frac{5}{8}$	$5\frac{1}{4}$	$4\frac{5}{8}$
20. MIRTH		$4\frac{7}{8}$	5	$4\frac{5}{8}$	$5\frac{1}{4}$	$4\frac{5}{16}$	$5\frac{1}{4}$	$4\frac{3}{4}$	$5\frac{1}{4}$	$4\frac{3}{8}$
21. IMITA.		$5\frac{1}{4}$	$5\frac{3}{8}$	$5\frac{1}{8}$	$5\frac{1}{4}$	$4\frac{7}{8}$	$5\frac{3}{8}$	5	$5\frac{5}{8}$	$5\frac{1}{16}$
22. INDIV.	$4\frac{5}{8}$	$4\frac{1}{4}$	$4\frac{3}{4}$	$4\frac{1}{4}$	$4\frac{1}{2}$	$4\frac{1}{16}$	5	$4\frac{1}{4}$	$4\frac{7}{8}$	$4\frac{1}{8}$
27. LOCAL.	$4\frac{5}{8}$	$4\frac{3}{8}$	5	$4\frac{1}{4}$	$4\frac{5}{8}$	$4\frac{1}{8}$	$5\frac{1}{8}$	$4\frac{1}{2}$	$5\frac{1}{4}$	$4\frac{3}{8}$
28. CALCU.		$3\frac{3}{4}$	$4\frac{1}{4}$	$3\frac{5}{8}$	$3\frac{7}{8}$	$3\frac{5}{8}$	$4\frac{1}{4}$	$3\frac{1}{2}$	$4\frac{1}{4}$	$3\frac{1}{2}$
29. ORDER	$4\frac{1}{4}$	4		4	$4\frac{1}{4}$		$4\frac{3}{8}$	$3\frac{7}{8}$		
30. EVENT.	$4\frac{3}{4}$	$4\frac{5}{8}$	5	$4\frac{7}{16}$	$4\frac{3}{4}$	$4\frac{1}{4}$	$5\frac{1}{4}$	$4\frac{1}{2}$	$5\frac{1}{4}$	$4\frac{3}{8}$
32. TUNE	$4\frac{1}{8}$	4	$4\frac{1}{8}$	$3\frac{3}{4}$	$3\frac{3}{8}$	$3\frac{3}{4}$	$4\frac{1}{2}$	$3\frac{1}{2}$	$4\frac{1}{8}$	$3\frac{5}{8}$
34. COMP.	$5\frac{3}{8}$	$5\frac{1}{8}$	$5\frac{1}{4}$	5	$5\frac{1}{4}$	$4\frac{5}{8}$	$5\frac{3}{8}$	5	$5\frac{5}{8}$	$4\frac{7}{8}$
35. CAUS.	$5\frac{1}{8}$	$5\frac{1}{8}$	$5\frac{1}{8}$	$4\frac{15}{16}$	$5\frac{1}{4}$	$4\frac{1}{2}$	$5\frac{1}{2}$	$4\frac{7}{8}$	$5\frac{1}{4}$	$4\frac{3}{4}$

Four Temperaments.
{ LYMPHATIC
  SANGUINE
  BILIOUS
  NERVOUS

Admeasurements of Heads of Distinguished Persons.	Gov. Lumpkin.	Belle of Market-st., Baltimore.	Late Hon. Felix Grundy.	The Beauty of Washington.	Hon. Mr. Graves.	The Charming Widow.	Hon. F. W. Pickens.	Miss B.	Gen. Scott.	Mrs. K.	Gen. Houston, of Texas.
1. AMAT.	$3\frac{1}{4}$	$2\frac{1}{2}$	$3\frac{1}{4}$	$2\frac{1}{4}$	$2\frac{1}{2}$	$2\frac{1}{4}$	$3\frac{1}{2}$	$2\frac{7}{8}$	$3\frac{3}{8}$	$2\frac{7}{8}$	$3\frac{1}{4}$
2. PHILO.	$4\frac{1}{2}$	$4\frac{1}{4}$	$4\frac{5}{8}$	4	$4\frac{1}{2}$		4	$3\frac{7}{8}$	$4\frac{3}{8}$	$3\frac{7}{8}$	$4\frac{3}{8}$
3. CONCE.	$5\frac{1}{4}$	$4\frac{3}{4}$	$5\frac{1}{4}$		$5\frac{1}{4}$		$4\frac{3}{4}$	$4\frac{1}{4}$	$5\frac{1}{4}$		$4\frac{3}{4}$
4. ADHES.		$3\frac{3}{4}$	4	4	$4\frac{3}{8}$	$4\frac{1}{8}$	$4\frac{1}{8}$	$4\frac{3}{8}$	$4\frac{3}{8}$	$4\frac{3}{8}$	$4\frac{1}{8}$
5. COMBA.	$3\frac{1}{4}$	$2\frac{7}{8}$	3	$2\frac{3}{4}$	3	$3\frac{1}{8}$	$3\frac{1}{8}$	$3\frac{1}{4}$	$3\frac{1}{2}$	$3\frac{1}{4}$	$3\frac{3}{8}$
6. DESTR.	$3\frac{1}{4}$	$3\frac{1}{8}$	$3\frac{1}{4}$	3	$3\frac{1}{4}$	$3\frac{1}{4}$	$3\frac{1}{4}$	3	$3\frac{1}{2}$	3	$3\frac{1}{8}$
ALIM.	$3\frac{1}{4}$	3		$2\frac{7}{8}$	$3\frac{1}{4}$			$2\frac{7}{8}$		$2\frac{7}{8}$	
7. SECRE.		$3\frac{5}{8}$	4	$3\frac{1}{4}$	$3\frac{3}{4}$			$3\frac{3}{4}$	$4\frac{1}{4}$	$3\frac{3}{4}$	$4\frac{1}{8}$
8. ACQUI.		$3\frac{1}{2}$		$3\frac{1}{4}$	4			$3\frac{3}{4}$	4	$3\frac{3}{4}$	
9. CONST.		$3\frac{3}{4}$	$3\frac{3}{4}$	$3\frac{1}{4}$	$4\frac{1}{4}$	$4\frac{1}{8}$	$4\frac{1}{8}$	$3\frac{5}{8}$	4	$3\frac{1}{2}$	$3\frac{1}{2}$
10. SELF-E.		$5\frac{1}{4}$	$5\frac{1}{4}$	5	$5\frac{3}{4}$	$5\frac{1}{4}$	$4\frac{3}{4}$	$5\frac{1}{4}$	$5\frac{3}{4}$	$5\frac{1}{4}$	$5\frac{3}{16}$
11. APPRO.	$5\frac{5}{8}$	$5\frac{1}{4}$	$5\frac{1}{4}$	5	$5\frac{1}{4}$		$5\frac{1}{2}$	$5\frac{1}{16}$	$5\frac{3}{4}$	$5\frac{1}{4}$	
12. CAU.	5	$4\frac{3}{4}$	$5\frac{1}{4}$	5	5	$4\frac{3}{4}$	$5\frac{1}{4}$	$4\frac{3}{4}$	$5\frac{1}{4}$	$4\frac{3}{4}$	$4\frac{7}{8}$
13. BENEV.	$5\frac{3}{4}$	$5\frac{1}{4}$	$5\frac{3}{4}$	$5\frac{5}{8}$	$5\frac{1}{2}$	$5\frac{5}{8}$	$5\frac{5}{8}$	$5\frac{1}{2}$	$5\frac{3}{4}$	$5\frac{1}{16}$	$5\frac{3}{4}$
14. VENER.	$5\frac{3}{4}$	$5\frac{3}{16}$	$5\frac{3}{16}$	$5\frac{3}{16}$	$5\frac{1}{2}$		$5\frac{5}{8}$	$5\frac{1}{2}$	$5\frac{7}{8}$	$5\frac{9}{16}$	$5\frac{1}{4}$
15. FIRM.	$5\frac{7}{8}$	$5\frac{1}{4}$	$5\frac{1}{16}$	$5\frac{1}{2}$	$5\frac{5}{8}$	$5\frac{9}{16}$	$5\frac{7}{8}$	$5\frac{1}{4}$	$6\frac{1}{16}$	$5\frac{5}{8}$	$5\frac{7}{8}$
16. CONSCI.		$5\frac{3}{4}$	$5\frac{7}{8}$	$5\frac{1}{4}$	$5\frac{1}{2}$		$5\frac{7}{8}$	$5\frac{1}{8}$	$5\frac{7}{8}$	$5\frac{1}{4}$	$5\frac{3}{4}$
17. HOPE			$5\frac{5}{8}$	5	$5\frac{1}{2}$		$5\frac{3}{4}$	$5\frac{1}{16}$		$5\frac{5}{8}$	
18. MARV.								$5\frac{1}{16}$		$5\frac{7}{16}$	$5\frac{5}{8}$
19. IDEAL.	5	$4\frac{1}{2}$	$5\frac{5}{8}$	5	$5\frac{1}{8}$	$5\frac{1}{4}$	$5\frac{1}{4}$	$4\frac{3}{4}$	$5\frac{1}{4}$	$4\frac{3}{4}$	5
20. MIRTH		$4\frac{1}{2}$	5	$4\frac{3}{4}$	$4\frac{7}{8}$	5	$5\frac{1}{4}$	$4\frac{7}{8}$	$5\frac{1}{4}$	5	$4\frac{7}{8}$
21. IMITA.		5	$5\frac{5}{8}$	$5\frac{1}{4}$	$5\frac{5}{8}$		$5\frac{5}{8}$	$5\frac{1}{16}$	$5\frac{5}{8}$	$5\frac{5}{8}$	$5\frac{1}{2}$
22. INDIV.	5	4	$4\frac{5}{8}$	$4\frac{3}{8}$	$4\frac{5}{8}$	$4\frac{5}{8}$	$4\frac{3}{4}$	$4\frac{1}{4}$	$4\frac{5}{8}$	$4\frac{1}{4}$	$4\frac{5}{8}$
27. LOCAL.	$4\frac{7}{8}$	$4\frac{1}{4}$	$4\frac{5}{8}$	$4\frac{3}{8}$	$4\frac{3}{8}$	$4\frac{3}{4}$		$4\frac{1}{2}$	$4\frac{3}{4}$	$4\frac{5}{8}$	
28. CALCU.	$4\frac{1}{4}$	$3\frac{1}{2}$	4	$3\frac{3}{8}$	$3\frac{3}{8}$	$3\frac{7}{8}$		$3\frac{3}{4}$	4	4	$3\frac{3}{4}$
29. ORDER	$4\frac{1}{2}$	$3\frac{1}{2}$	$4\frac{1}{4}$	4	$4\frac{1}{4}$			$4\frac{1}{16}$	$4\frac{3}{8}$	$4\frac{1}{4}$	$4\frac{1}{16}$
30. EVENT.	$4\frac{5}{16}$	$4\frac{3}{4}$	$4\frac{1}{4}$	$4\frac{5}{8}$		5	$4\frac{5}{8}$	$4\frac{9}{16}$	5	$4\frac{3}{8}$	5
32. TUNE	$4\frac{1}{8}$		4	$3\frac{9}{16}$	$3\frac{7}{8}$	4	$4\frac{1}{8}$	4	4	$3\frac{7}{8}$	
34. COMP	$5\frac{1}{4}$	$4\frac{3}{4}$	$5\frac{3}{8}$	5	$5\frac{3}{16}$	$5\frac{5}{8}$	$5\frac{1}{4}$	$4\frac{1}{16}$	$5\frac{5}{8}$	$5\frac{1}{4}$	$5\frac{5}{8}$
35. CAUS	$5\frac{3}{8}$	$4\frac{3}{4}$	$5\frac{3}{16}$	$4\frac{7}{8}$	$5\frac{3}{16}$	$5\frac{1}{4}$	$5\frac{1}{4}$	$4\frac{7}{8}$	$5\frac{1}{4}$	$5\frac{1}{8}$	$5\frac{1}{2}$

Admeasurements of Heads of Distinguished Persons.	Rev. Dr. Channing.	Rev. Mr. Pierpont.	Dr. Warren, the Anatomist.	Mr. Blair, Editor of the Globe.	Mr. F., of N. Y., Prest. St. G. S.	Dr. Jones, Washington.	W. L. Garrison, Abolitionist.	E. Quincy, Abolitionist.	Mr. Porter, the Kentucky Giant.	Major Stevens, Am. Dwarf.	Monsieur Bihin, Belgian Giant.
1. AMAT.	3	$2\frac{7}{8}$	$2\frac{5}{8}$	$2\frac{1}{2}$	$2\frac{1}{2}$	$2\frac{3}{4}$	$2\frac{1}{2}$	3	3	$2\frac{5}{8}$	$3\frac{1}{2}$
2. PHILO.	4	$4\frac{1}{4}$	$4\frac{5}{8}$	4	$3\frac{3}{4}$	4	$3\frac{5}{8}$	$4\frac{1}{4}$	$3\frac{7}{8}$	$2\frac{3}{4}$	$4\frac{1}{8}$
3. CONCE.	$4\frac{1}{8}$	$4\frac{7}{8}$	$4\frac{7}{8}$	$4\frac{3}{4}$	$4\frac{3}{4}$	5			$5\frac{5}{8}$	4	$5\frac{1}{8}$
4. ADHES.	$3\frac{7}{8}$	$3\frac{7}{8}$	$4\frac{3}{16}$	$3\frac{3}{8}$	$3\frac{5}{8}$	4	$3\frac{5}{8}$		$4\frac{5}{8}$	$3\frac{1}{4}$	$4\frac{3}{8}$
5. COMBA.	$3\frac{3}{8}$	$3\frac{3}{8}$	$3\frac{5}{8}$	$2\frac{3}{4}$	$2\frac{7}{8}$	$2\frac{7}{8}$	$2\frac{3}{4}$	$3\frac{3}{8}$	$3\frac{1}{2}$	$3\frac{1}{8}$	$3\frac{5}{8}$
6. DESTR.	$3\frac{1}{16}$	$3\frac{3}{16}$	$3\frac{1}{4}$	3	3	$2\frac{3}{4}$	$2\frac{7}{8}$	$3\frac{1}{8}$	$2\frac{7}{8}$	$2\frac{7}{8}$	$3\frac{1}{2}$
ALIM.						$2\frac{3}{4}$					$3\frac{1}{2}$
7. SECRE.	$3\frac{3}{4}$	$4\frac{3}{16}$	$4\frac{1}{4}$	$3\frac{1}{2}$	$3\frac{5}{8}$	$3\frac{1}{4}$			$3\frac{7}{8}$	$3\frac{3}{4}$	$4\frac{1}{4}$
8. ACQUI.	4	$3\frac{3}{4}$	$4\frac{1}{4}$	4	$3\frac{3}{4}$	$3\frac{5}{8}$			$3\frac{3}{4}$	4	$4\frac{3}{8}$
9. CONST.	$3\frac{7}{8}$	$3\frac{3}{4}$		4	$3\frac{1}{2}$	$3\frac{3}{4}$			$3\frac{3}{8}$	$3\frac{5}{8}$	$4\frac{1}{4}$
10. SELF-E.	5	$5\frac{5}{8}$	$5\frac{5}{8}$	$5\frac{3}{8}$	5	$5\frac{1}{2}$	$5\frac{3}{8}$	$5\frac{5}{8}$	$6\frac{1}{8}$	$5\frac{3}{16}$	6
11. APPRO.	$5\frac{1}{16}$	$5\frac{1}{2}$	$5\frac{1}{2}$	$5\frac{1}{8}$	5	$5\frac{3}{8}$	5	$5\frac{3}{8}$	6	$5\frac{1}{16}$	$5\frac{7}{8}$
12. CAU.	$4\frac{5}{8}$	$5\frac{1}{8}$	$5\frac{1}{8}$	$4\frac{3}{8}$	$4\frac{7}{8}$	5	$4\frac{3}{8}$	$5\frac{1}{4}$	$5\frac{1}{8}$	$4\frac{6}{8}$	$5\frac{3}{8}$
13. BENEV.	$5\frac{3}{8}$	$5\frac{11}{16}$	$5\frac{3}{4}$	$5\frac{5}{8}$	$5\frac{3}{4}$	$5\frac{1}{2}$	$5\frac{5}{8}$	$5\frac{3}{4}$	$6\frac{1}{8}$	$5\frac{1}{4}$	6
14. VENER.	$5\frac{3}{8}$	$5\frac{13}{16}$	$5\frac{3}{4}$	$5\frac{5}{8}$	$5\frac{3}{4}$	$5\frac{5}{8}$	$5\frac{1}{2}$	$5\frac{11}{16}$	$6\frac{5}{16}$	$5\frac{3}{8}$	$6\frac{1}{4}$
15. FIRM.	$5\frac{1}{2}$	$5\frac{13}{16}$	$5\frac{13}{16}$	$5\frac{3}{4}$	$5\frac{3}{4}$	$5\frac{3}{4}$	$5\frac{5}{8}$	$5\frac{3}{4}$	$6\frac{3}{8}$	$5\frac{3}{8}$	$6\frac{1}{4}$
16. CONSCI.	$5\frac{1}{4}$	$5\frac{11}{16}$	$5\frac{11}{16}$	$5\frac{3}{8}$	$5\frac{5}{8}$	$5\frac{5}{8}$			$6\frac{5}{16}$	$5\frac{1}{4}$	$6\frac{1}{8}$
17. HOPE	$5\frac{1}{4}$	$5\frac{9}{16}$	$5\frac{11}{16}$	$5\frac{3}{8}$	$5\frac{3}{8}$	$5\frac{1}{2}$			$6\frac{1}{4}$	$5\frac{1}{8}$	6
18. MARV.	$5\frac{1}{8}$	$5\frac{1}{2}$	$5\frac{3}{8}$								
19. IDEAL.	$4\frac{5}{8}$	$5\frac{3}{16}$	$5\frac{1}{8}$	5	$5\frac{1}{4}$	$5\frac{1}{8}$	$4\frac{3}{4}$	$4\frac{7}{8}$	$5\frac{3}{8}$	$4\frac{12}{16}$	$5\frac{5}{8}$
20. MIRTH	$4\frac{5}{8}$	5	$5\frac{1}{4}$	$5\frac{1}{8}$	$4\frac{7}{8}$	5			$5\frac{3}{8}$	$4\frac{5}{8}$	$5\frac{3}{8}$
21. IMITA.	$5\frac{1}{8}$	$5\frac{1}{2}$	$5\frac{3}{8}$		$5\frac{1}{4}$	$5\frac{1}{2}$			$5\frac{7}{8}$	$5\frac{1}{16}$	6
22. INDIV.	$4\frac{1}{4}$	$4\frac{1}{8}$	5	$4\frac{5}{8}$	$4\frac{1}{2}$	$4\frac{5}{8}$	$4\frac{5}{8}$	$4\frac{5}{8}$	$4\frac{7}{8}$	4	$4\frac{1}{2}$
27. LOCAL.	$4\frac{3}{8}$	$4\frac{3}{8}$	$5\frac{1}{8}$		$4\frac{5}{8}$	$4\frac{3}{4}$		$4\frac{5}{8}$	5	$4\frac{1}{8}$	$4\frac{3}{4}$
28. CALCU.	$3\frac{3}{4}$	$3\frac{3}{4}$	4	$4\frac{5}{8}$		$3\frac{3}{4}$	4	$4\frac{1}{4}$	$4\frac{1}{4}$	$3\frac{9}{16}$	$4\frac{3}{8}$
29. ORDER	4		$4\frac{3}{4}$	$4\frac{3}{8}$		$4\frac{1}{8}$		$4\frac{1}{2}$	$4\frac{1}{2}$	$3\frac{11}{16}$	$4\frac{1}{2}$
30. EVENT.	$4\frac{5}{8}$		$5\frac{1}{4}$	5	$4\frac{5}{8}$				$5\frac{1}{8}$	$4\frac{5}{8}$	
32. TUNE				$4\frac{1}{4}$	$3\frac{1}{2}$	$3\frac{5}{8}$			$4\frac{1}{2}$	$3\frac{5}{8}$	$4\frac{1}{4}$
34. COMP.	5	$5\frac{3}{16}$	$5\frac{1}{2}$	$5\frac{3}{8}$	$5\frac{5}{8}$	$5\frac{3}{16}$	$5\frac{3}{8}$	$5\frac{3}{8}$	$5\frac{3}{8}$	$4\frac{3}{4}$	$5\frac{1}{2}$
35. CAUS.	5	$5\frac{1}{8}$	$5\frac{1}{2}$	$5\frac{3}{16}$	$5\frac{5}{16}$	$5\frac{1}{16}$	$5\frac{1}{4}$	$5\frac{1}{4}$	$5\frac{3}{8}$	$4\frac{9}{16}$	$5\frac{1}{2}$

Admeasurements of Heads of Remarkable Persons.	Charles Freeman, American Giant.	Louisville Dwarf.	Dr. Valentine, Humorist.	Deaf Burke, the Pugilist.	Mr. O'Rourke, Pugilist.	Charles Lovell, 6 years old.	Idiot, 9 yrs. old.	Infant of 18 mos.	Murderer.	Boston Boy, 10 years old.	Negro Murderer.
1. Amat.	$3\frac{1}{4}$	$2\frac{1}{2}$	$2\frac{1}{2}$	$3\frac{3}{4}$	$3\frac{3}{8}$	2	$2\frac{1}{4}$	2	$3\frac{5}{8}$	$3\frac{5}{8}$	$4\frac{3}{8}$
2. Philo.	$4\frac{7}{8}$	$3\frac{3}{8}$	$3\frac{3}{8}$	$4\frac{1}{4}$	4	$4\frac{1}{16}$	$3\frac{5}{8}$	$3\frac{3}{4}$	$4\frac{1}{4}$	$4\frac{1}{8}$	$5\frac{1}{16}$
3. Conce.	$5\frac{3}{8}$	$4\frac{5}{8}$	$4\frac{3}{4}$	5	$4\frac{3}{8}$	$4\frac{7}{16}$			$5\frac{1}{2}$		$5\frac{1}{2}$
4. Adhes.	$4\frac{1}{4}$	$3\frac{1}{2}$	$3\frac{1}{4}$	$4\frac{1}{4}$	4	$3\frac{3}{8}$	$3\frac{1}{4}$	$3\frac{1}{8}$	$4\frac{1}{4}$	$4\frac{3}{8}$	$5\frac{1}{8}$
5. Comba	$3\frac{1}{2}$	$3\frac{1}{4}$	3	$3\frac{7}{8}$	$3\frac{5}{16}$	$3\frac{9}{16}$	$2\frac{1}{16}$	$2\frac{1}{4}$	$3\frac{7}{8}$	$3\frac{5}{8}$	$4\frac{1}{8}$
6. Destr.	$3\frac{1}{2}$	$3\frac{1}{4}$	$3\frac{1}{8}$	$3\frac{3}{4}$	$3\frac{1}{2}$	$3\frac{5}{8}$	$2\frac{5}{8}$	$2\frac{5}{8}$	$3\frac{5}{8}$	$3\frac{5}{16}$	$3\frac{9}{16}$
Alim.											
7. Secre.	$4\frac{1}{8}$	4	$3\frac{1}{2}$	$3\frac{5}{8}$	$4\frac{1}{4}$	$3\frac{7}{8}$	$3\frac{3}{8}$		$4\frac{1}{8}$	$4\frac{3}{8}$	$4\frac{3}{8}$
8. Acqui.		4	$3\frac{3}{4}$	$3\frac{1}{2}$		$3\frac{7}{8}$			$4\frac{1}{8}$	4	$4\frac{3}{8}$
9. Const.	$3\frac{3}{4}$	$3\frac{3}{4}$	4	$3\frac{7}{8}$	$4\frac{1}{4}$	$3\frac{1}{2}$	$3\frac{1}{8}$	$3\frac{3}{8}$	$4\frac{3}{8}$	4	4
10. Self-E.	$5\frac{1}{4}$	$5\frac{3}{8}$	$5\frac{1}{4}$	4	$5\frac{1}{4}$	5	$4\frac{3}{4}$	$4\frac{5}{8}$	$5\frac{1}{2}$	$6\frac{1}{8}$	$5\frac{1}{4}$
11. Appro.	$5\frac{1}{2}$	$5\frac{1}{8}$	$5\frac{5}{8}$	$5\frac{7}{8}$		5	$4\frac{1}{2}$	$4\frac{1}{2}$	$5\frac{3}{8}$	6	$5\frac{3}{8}$
12. Cau.	$5\frac{1}{8}$	$4\frac{7}{8}$	$4\frac{3}{4}$	$5\frac{1}{8}$	$5\frac{1}{8}$	$5\frac{5}{16}$	$4\frac{3}{8}$	4	$4\frac{7}{8}$	$5\frac{1}{4}$	5
13. Benev.	$5\frac{3}{16}$	$5\frac{3}{8}$	$5\frac{1}{2}$	$4\frac{1}{4}$	$5\frac{3}{8}$	$5\frac{9}{16}$	$4\frac{7}{8}$	$4\frac{7}{8}$	$5\frac{3}{8}$	$6\frac{1}{4}$	$5\frac{3}{4}$
14. Vener.	$5\frac{3}{4}$	$5\frac{1}{2}$	$5\frac{5}{8}$	$5\frac{3}{8}$	$5\frac{5}{8}$	$5\frac{12}{16}$	$4\frac{7}{8}$	5	$5\frac{3}{8}$	$6\frac{1}{4}$	$5\frac{7}{8}$
15. Firm.	6	$5\frac{5}{8}$	$5\frac{3}{8}$	$5\frac{7}{8}$	6	$5\frac{5}{8}$	5	$5\frac{1}{8}$	$5\frac{7}{16}$	$6\frac{1}{8}$	$5\frac{7}{8}$
16. Consci.	$5\frac{3}{4}$	$5\frac{5}{8}$	$5\frac{5}{8}$	6		$5\frac{5}{8}$	$4\frac{7}{8}$	$4\frac{13}{16}$	$5\frac{5}{16}$		$5\frac{3}{4}$
17. Hope	$5\frac{5}{8}$	$5\frac{5}{8}$	$5\frac{5}{8}$	$5\frac{3}{4}$		$5\frac{1}{12}$			$5\frac{1}{4}$		$5\frac{11}{16}$
18. Marv.						$5\frac{1}{2}$					
19. Ideal.	5	5	5	$5\frac{5}{8}$	$5\frac{3}{8}$	$4\frac{3}{8}$	$4\frac{3}{8}$	$4\frac{1}{2}$	$5\frac{1}{8}$	$5\frac{1}{2}$	5
20. Mirth	$4\frac{7}{8}$	$4\frac{3}{4}$	$4\frac{7}{8}$	$4\frac{7}{8}$	$4\frac{1}{2}$	$4\frac{1}{2}$	$4\frac{1}{4}$		$5\frac{3}{16}$		$5\frac{1}{4}$
21. Imita.	$5\frac{1}{2}$	$5\frac{5}{8}$	$5\frac{5}{8}$	$5\frac{3}{4}$		$5\frac{1}{4}$	$4\frac{1}{2}$	$4\frac{3}{8}$	$5\frac{3}{8}$	$6\frac{1}{4}$	$5\frac{1}{2}$
22. Indiv.	$4\frac{5}{8}$	$4\frac{5}{16}$	$4\frac{1}{2}$	5	$4\frac{5}{8}$	$4\frac{1}{8}$	4	$3\frac{5}{8}$	$4\frac{1}{4}$	$4\frac{3}{8}$	$4\frac{7}{8}$
27. Local.	$4\frac{1}{4}$	$4\frac{3}{8}$	$4\frac{5}{8}$	$3\frac{5}{8}$	$4\frac{7}{8}$	$4\frac{2}{8}$	4	$3\frac{3}{8}$	$5\frac{1}{4}$		5
28. Calcu.	4	4			4	$3\frac{7}{8}$	$3\frac{3}{4}$	$3\frac{3}{8}$	$4\frac{1}{8}$		$4\frac{3}{8}$
29. Order	$4\frac{1}{4}$	$3\frac{3}{4}$	$4\frac{1}{8}$	$4\frac{1}{2}$		$4\frac{1}{2}$	$3\frac{9}{16}$	$3\frac{1}{4}$			$4\frac{3}{8}$
30. Event.	$4\frac{3}{8}$	$4\frac{3}{8}$	$4\frac{3}{4}$			$4\frac{1}{2}$			$5\frac{3}{8}$		$5\frac{1}{16}$
32. Tune		$3\frac{7}{8}$	4	$4\frac{3}{4}$	$4\frac{1}{4}$		$3\frac{1}{2}$		$4\frac{3}{8}$		
34. Comp.	$5\frac{1}{4}$	5	$5\frac{1}{4}$	5	$5\frac{3}{8}$	$4\frac{7}{8}$	$4\frac{3}{8}$	$4\frac{1}{4}$	$5\frac{3}{8}$	$5\frac{1}{2}$	$5\frac{3}{8}$
35. Caus.	$5\frac{1}{8}$	$4\frac{7}{8}$	$5\frac{1}{8}$	$4\frac{7}{8}$	$5\frac{1}{4}$	$4\frac{7}{8}$	$4\frac{3}{16}$	$4\frac{1}{4}$	$5\frac{1}{8}$	$5\frac{1}{2}$	$5\frac{1}{4}$

Admeasurements of Skulls.	Spurzheim's Skull; his own bequest.	Egypt. Mummy.	Peruvian Child.	Tattooed N. Zealand Chief.	Hindoo Male.	Hindoo Female.	Male Hindoo.	Female Hindoo.	Peruvian.	Abyssinian black	Malay.
1. AMAT.	$3\frac{5}{8}$	$3\frac{1}{4}$	$2\frac{1}{2}$	$3\frac{3}{8}$	$3\frac{1}{8}$	3	$3\frac{1}{4}$	$3\frac{1}{4}$	$2\frac{3}{4}$	$3\frac{1}{4}$	$3\frac{1}{16}$
2. PHILO.	4	$4\frac{1}{4}$	$3\frac{1}{2}$	$3\frac{7}{8}$	$3\frac{7}{8}$	$3\frac{7}{8}$	$3\frac{7}{8}$	$3\frac{5}{8}$	$3\frac{5}{8}$	$3\frac{3}{8}$	$4\frac{1}{2}$
3. CONCE.	$4\frac{1}{2}$	$4\frac{1}{2}$	$3\frac{3}{4}$	$4\frac{1}{4}$	$4\frac{5}{8}$	$3\frac{5}{8}$	$4\frac{3}{4}$	$4\frac{1}{8}$	$4\frac{1}{2}$	$4\frac{1}{2}$	
4. ADHES.	$3\frac{7}{8}$	4	$2\frac{7}{8}$	$3\frac{1}{4}$	$3\frac{7}{8}$	$3\frac{1}{2}$	$3\frac{7}{8}$	$3\frac{9}{16}$		$3\frac{5}{8}$	$4\frac{3}{8}$
5. COMBA.	$3\frac{5}{8}$	$2\frac{7}{8}$	$2\frac{3}{4}$	$2\frac{1}{2}$	3	$2\frac{5}{8}$	3	$2\frac{3}{8}$	$2\frac{3}{4}$	$2\frac{5}{8}$	3
6. DESTR.	3	$2\frac{4}{16}$	$2\frac{3}{16}$	$2\frac{1}{4}$	$2\frac{3}{8}$	$2\frac{5}{8}$	$2\frac{1}{2}$	$2\frac{1}{8}$	$2\frac{1}{2}$	$2\frac{1}{4}$	$3\frac{3}{8}$
ALIM.	$2\frac{3}{4}$	$2\frac{1}{16}$	$2\frac{1}{8}$						$2\frac{3}{8}$	2	
7. SECRE.	$3\frac{1}{2}$	$3\frac{1}{4}$	$3\frac{1}{4}$	3	$3\frac{1}{8}$	3	$3\frac{3}{8}$	$2\frac{5}{8}$	$3\frac{1}{4}$	$3\frac{1}{4}$	$3\frac{3}{8}$
8. ACQUI.	$3\frac{5}{8}$	$3\frac{1}{8}$	$3\frac{1}{4}$	$3\frac{1}{8}$	$3\frac{3}{8}$	$3\frac{1}{4}$	$3\frac{1}{2}$	3	$3\frac{3}{8}$	$3\frac{1}{4}$	
9. CONST.	$3\frac{5}{8}$	$3\frac{1}{4}$	$3\frac{1}{4}$	$3\frac{1}{8}$	$3\frac{5}{8}$	$3\frac{1}{4}$	$3\frac{1}{2}$	$3\frac{1}{8}$	$3\frac{5}{8}$	$3\frac{1}{4}$	$3\frac{1}{2}$
10. SELF-E.	5	$4\frac{5}{8}$	$4\frac{1}{4}$	$4\frac{5}{8}$	$4\frac{7}{8}$	$4\frac{9}{16}$	$4\frac{7}{8}$	$4\frac{5}{8}$	$5\frac{1}{16}$	$4\frac{3}{8}$	$5\frac{3}{16}$
11. APPRO.	$4\frac{7}{8}$	$4\frac{3}{8}$	$1\frac{9}{16}$	$4\frac{1}{2}$	$4\frac{7}{8}$	$4\frac{1}{2}$	$4\frac{7}{8}$	$4\frac{1}{2}$	$4\frac{7}{8}$	$4\frac{3}{8}$	$4\frac{7}{8}$
12. CAU.	$4\frac{9}{16}$	$4\frac{1}{16}$	$4\frac{1}{2}$	$4\frac{1}{16}$	$4\frac{9}{16}$	$4\frac{1}{16}$	$4\frac{3}{8}$	4	$4\frac{3}{8}$	$4\frac{3}{8}$	$4\frac{3}{8}$
13. BENEV.	$5\frac{1}{16}$	$4\frac{1}{2}$	$4\frac{5}{16}$	$4\frac{3}{8}$	5	$4\frac{9}{16}$	5	$4\frac{1}{2}$	$4\frac{7}{8}$	$4\frac{9}{16}$	$4\frac{1}{2}$
14. VENER.	$5\frac{1}{16}$	$4\frac{9}{16}$	$4\frac{1}{2}$	$4\frac{3}{8}$	$5\frac{1}{4}$	$4\frac{9}{16}$	$4\frac{7}{8}$	$4\frac{9}{16}$	5	$4\frac{5}{8}$	$4\frac{13}{16}$
15. FIRM.	$5\frac{1}{8}$	$4\frac{11}{16}$	$4\frac{5}{8}$	$4\frac{13}{16}$	$5\frac{1}{4}$	$4\frac{5}{8}$	$5\frac{1}{16}$	$4\frac{5}{8}$	$5\frac{1}{8}$	$4\frac{11}{16}$	5
16. CONSCI.	$5\frac{1}{16}$	$4\frac{11}{16}$	$4\frac{5}{8}$	$4\frac{11}{16}$	$5\frac{1}{16}$	$4\frac{9}{16}$	5	$4\frac{9}{16}$	5	$4\frac{11}{16}$	$4\frac{7}{8}$
17. HOPE	$5\frac{7}{16}$	$4\frac{7}{16}$	$4\frac{1}{2}$	$4\frac{5}{8}$	5	$4\frac{1}{2}$	$4\frac{7}{8}$	$4\frac{3}{8}$	$4\frac{7}{8}$	$4\frac{5}{8}$	$4\frac{11}{16}$
18. MARV.	$4\frac{7}{8}$	$4\frac{1}{16}$	$4\frac{7}{16}$						$4\frac{7}{8}$	$4\frac{9}{16}$	$4\frac{5}{8}$
19. IDEAL.	$4\frac{11}{16}$	$4\frac{1}{2}$	$4\frac{1}{8}$	$4\frac{3}{8}$	$4\frac{9}{16}$	$4\frac{3}{16}$	$4\frac{7}{16}$	$4\frac{1}{8}$	$4\frac{3}{8}$	$4\frac{3}{16}$	$4\frac{3}{8}$
20. MIRTH	$4\frac{5}{8}$	$4\frac{3}{16}$	$3\frac{7}{8}$	$4\frac{3}{8}$	$4\frac{5}{8}$	$4\frac{1}{4}$	$4\frac{4}{8}$	$4\frac{3}{16}$	4	$4\frac{1}{8}$	$4\frac{5}{16}$
21. IMITA.	$4\frac{7}{8}$	$4\frac{3}{8}$	$4\frac{1}{4}$	$4\frac{11}{16}$	$4\frac{7}{8}$	$4\frac{1}{4}$	$4\frac{3}{16}$	$4\frac{3}{8}$	$4\frac{5}{8}$	$4\frac{3}{8}$	$4\frac{7}{16}$
22. INDIV.	$4\frac{3}{8}$	$3\frac{7}{8}$	$3\frac{1}{2}$	$4\frac{1}{4}$	$4\frac{5}{16}$	$3\frac{3}{4}$	$3\frac{7}{8}$	$3\frac{7}{8}$	4	$3\frac{3}{16}$	$4\frac{1}{16}$
27. LOCAL.	$4\frac{7}{16}$	$3\frac{15}{16}$	$3\frac{5}{8}$	$4\frac{5}{16}$	$4\frac{3}{8}$	$3\frac{15}{16}$	$4\frac{1}{4}$	$3\frac{7}{8}$	$4\frac{1}{8}$	$3\frac{7}{8}$	$4\frac{7}{16}$
28. CALCU.	$3\frac{7}{8}$	$3\frac{3}{8}$	$3\frac{1}{4}$	4	$3\frac{3}{4}$	$3\frac{3}{8}$	$3\frac{5}{8}$	$3\frac{5}{16}$	$3\frac{5}{8}$	$3\frac{1}{2}$	$3\frac{5}{8}$
29. ORDER	$4\frac{1}{4}$	$3\frac{11}{16}$	$3\frac{3}{4}$	$4\frac{7}{16}$	4	$3\frac{1}{2}$	$3\frac{3}{8}$	$3\frac{1}{2}$	$3\frac{3}{8}$	$3\frac{5}{8}$	$3\frac{7}{8}$
30. EVENT.	$4\frac{1}{2}$	4	$3\frac{11}{16}$	$4\frac{3}{8}$	$4\frac{7}{16}$	4	$4\frac{1}{4}$	4	$4\frac{1}{4}$	4	$4\frac{1}{4}$
32. TUNE	$3\frac{7}{8}$	$3\frac{1}{2}$	$3\frac{1}{8}$						$3\frac{3}{8}$	$3\frac{1}{2}$	$3\frac{5}{8}$
34. COMP.	$4\frac{7}{8}$	$4\frac{1}{4}$	$3\frac{15}{16}$	$4\frac{9}{16}$	$4\frac{3}{4}$	$4\frac{1}{4}$	$4\frac{9}{16}$	$4\frac{5}{16}$	$4\frac{1}{2}$	$4\frac{3}{16}$	$4\frac{3}{8}$
35. CAUS.	$4\frac{13}{16}$	$4\frac{1}{4}$	4	$4\frac{7}{16}$	$4\frac{9}{16}$	$4\frac{3}{8}$	$4\frac{9}{16}$	$4\frac{3}{16}$	$4\frac{7}{16}$	$4\frac{1}{8}$	$4\frac{3}{8}$

Admeasurements of Skulls.	Turk.	Austrian.	Greek Female *	Greek Male.*	Irish Gravedigger.*	Spanish Pirate.*	Maccasser.	Roman from the Catacombs.	Sandwich Islander.	Javanese.	New Hollander.
1. AMAT.	$3\frac{5}{8}$	$3\frac{3}{4}$	$3\frac{3}{8}$	$3\frac{3}{8}$	$3\frac{7}{8}$	$3\frac{5}{8}$	$3\frac{1}{4}$	$3\frac{7}{8}$	$3\frac{1}{16}$	$3\frac{1}{12}$	$3\frac{1}{12}$
2. PHILO.	$4\frac{1}{8}$	$4\frac{1}{2}$	$3\frac{7}{8}$	$4\frac{1}{4}$	$4\frac{5}{8}$	$4\frac{1}{8}$	$4\frac{1}{8}$	$4\frac{3}{16}$	4	$4\frac{1}{4}$	$4\frac{1}{8}$
3. CONCE.	$4\frac{3}{8}$	$5\frac{1}{8}$	$4\frac{1}{8}$	$4\frac{5}{8}$	$4\frac{6}{8}$	$4\frac{1}{2}$	$4\frac{1}{2}$	$4\frac{5}{16}$	$4\frac{1}{2}$	$4\frac{1}{16}$	$4\frac{1}{4}$
4. ADHES.	$3\frac{7}{8}$	$4\frac{3}{8}$	$3\frac{5}{8}$	4	$4\frac{1}{4}$	4	4	$3\frac{7}{8}$	$4\frac{1}{16}$	$3\frac{5}{8}$	$4\frac{1}{4}$
5. COMBA.	3	$3\frac{1}{4}$	$3\frac{3}{8}$	3	$3\frac{4}{16}$	$3\frac{1}{8}$	3	$3\frac{7}{8}$	3	3	$3\frac{1}{8}$
6. DESTR.	$2\frac{1}{4}$	$2\frac{3}{8}$	$2\frac{1}{4}$	$2\frac{1}{4}$	$2\frac{3}{8}$	$2\frac{5}{8}$	$2\frac{3}{8}$	$3\frac{3}{8}$	$3\frac{1}{2}$	$2\frac{3}{8}$	$2\frac{1}{4}$
ALIM.	$2\frac{1}{8}$	$2\frac{3}{16}$					$2\frac{1}{4}$	$3\frac{1}{8}$	$3\frac{3}{8}$	$2\frac{1}{4}$	$2\frac{9}{16}$
7. SECRE.	$3\frac{1}{16}$	$3\frac{1}{4}$	3	3	$3\frac{3}{8}$	$3\frac{1}{2}$	$3\frac{3}{16}$	$3\frac{1}{16}$	$3\frac{1}{4}$	$3\frac{3}{16}$	$3\frac{1}{4}$
8. ACQUI.	$3\frac{1}{4}$	$3\frac{1}{2}$	$3\frac{1}{8}$	$3\frac{3}{8}$	$3\frac{7}{16}$	$3\frac{5}{8}$	$3\frac{3}{8}$	$3\frac{3}{8}$	$3\frac{3}{8}$	$3\frac{3}{16}$	$3\frac{1}{4}$
9. CONST.	$3\frac{1}{16}$	$3\frac{1}{2}$	$3\frac{1}{8}$	$3\frac{1}{4}$	$3\frac{1}{2}$	$3\frac{5}{8}$	$3\frac{3}{8}$	$3\frac{3}{8}$	$3\frac{3}{8}$	$3\frac{3}{4}$	$3\frac{1}{2}$
10. SELF-E.	$4\frac{1}{2}$	$5\frac{1}{4}$	$4\frac{7}{16}$	$4\frac{7}{8}$	$4\frac{7}{8}$	$4\frac{7}{8}$	$4\frac{3}{4}$	$4\frac{11}{16}$	$4\frac{13}{16}$	$5\frac{1}{16}$	$5\frac{1}{4}$
11. APPRO.	$4\frac{7}{16}$	$5\frac{1}{8}$	$4\frac{1}{16}$	$4\frac{3}{8}$	5	$4\frac{6}{8}$	$4\frac{1}{16}$	$4\frac{9}{16}$	$4\frac{3}{16}$	$4\frac{7}{16}$	$5\frac{1}{4}$
12. CAU.	$3\frac{3}{4}$	$4\frac{9}{16}$	$3\frac{7}{8}$	$4\frac{1}{4}$	$4\frac{3}{8}$	$4\frac{4}{8}$	$4\frac{1}{4}$	$4\frac{1}{16}$	$4\frac{1}{4}$	$4\frac{3}{8}$	$4\frac{3}{4}$
13. BENEV.	$4\frac{1}{2}$	$5\frac{1}{16}$	$4\frac{1}{2}$	$4\frac{1}{16}$	$4\frac{7}{8}$	$4\frac{7}{8}$	$4\frac{3}{4}$	$4\frac{1}{2}$	$4\frac{3}{16}$	$4\frac{3}{16}$	$5\frac{1}{16}$
14. VENER.	$4\frac{9}{16}$	$5\frac{1}{16}$	$4\frac{7}{16}$	$4\frac{3}{8}$	5	$5\frac{1}{16}$	$4\frac{3}{16}$	$4\frac{1}{2}$	$4\frac{5}{16}$	$4\frac{3}{16}$	$5\frac{3}{16}$
15. FIRM.	$4\frac{5}{8}$	$5\frac{3}{16}$	$4\frac{1}{2}$	$4\frac{7}{8}$	$5\frac{1}{16}$	$5\frac{1}{4}$	$4\frac{7}{8}$	$4\frac{5}{8}$	5	5	$5\frac{1}{4}$
16. CONSCI.	$4\frac{1}{4}$	$5\frac{1}{16}$	$4\frac{3}{8}$	$4\frac{3}{8}$	$4\frac{1}{16}$	$4\frac{5}{16}$	$4\frac{1}{16}$	$4\frac{1}{16}$	$4\frac{3}{16}$	$4\frac{7}{8}$	$5\frac{3}{16}$
17. HOPE	$4\frac{3}{8}$	$5\frac{1}{16}$	$4\frac{5}{16}$	$4\frac{1}{16}$	$4\frac{7}{8}$	$4\frac{3}{16}$	$4\frac{1}{16}$	$4\frac{9}{16}$	$4\frac{3}{16}$	$4\frac{3}{16}$	$5\frac{1}{16}$
18. MARV.	$4\frac{1}{4}$	5					$4\frac{9}{16}$	$4\frac{1}{2}$	$4\frac{3}{4}$	$4\frac{3}{4}$	$4\frac{7}{8}$
19. IDEAL.	$4\frac{1}{8}$	$4\frac{5}{8}$	$4\frac{1}{16}$	$4\frac{3}{16}$	$4\frac{5}{16}$	$4\frac{7}{16}$	$4\frac{5}{16}$	$4\frac{5}{16}$	$4\frac{3}{8}$	$4\frac{9}{16}$	$4\frac{5}{8}$
20. MIRTH	$4\frac{1}{8}$	$4\frac{3}{4}$	$4\frac{1}{4}$	$4\frac{5}{16}$	$4\frac{9}{16}$	$4\frac{1}{2}$	$4\frac{3}{8}$	$4\frac{1}{4}$	$4\frac{5}{16}$	$4\frac{7}{16}$	$4\frac{5}{8}$
21. IMITA.	$4\frac{3}{8}$	5	$4\frac{3}{8}$	$4\frac{5}{8}$	$4\frac{13}{16}$	$4\frac{13}{16}$	$4\frac{1}{4}$	$4\frac{7}{16}$	$4\frac{5}{8}$	$4\frac{11}{16}$	5
22. INDIV.	4	$4\frac{3}{16}$	$3\frac{7}{8}$	4	$3\frac{15}{16}$	$4\frac{5}{16}$	$4\frac{1}{4}$	$3\frac{11}{16}$	$3\frac{3}{16}$	$4\frac{1}{8}$	$4\frac{7}{16}$
27. LOCAL.	$4\frac{1}{16}$	$4\frac{5}{16}$	4	$4\frac{1}{16}$	$4\frac{1}{16}$	$4\frac{3}{8}$	$4\frac{1}{4}$	$4\frac{3}{16}$	$4\frac{1}{4}$	$4\frac{3}{16}$	$4\frac{9}{16}$
28. CALCU.	$3\frac{1}{4}$	$3\frac{11}{16}$	$3\frac{5}{8}$	$3\frac{9}{16}$	$3\frac{1}{2}$	$4\frac{1}{8}$	$3\frac{7}{8}$	$3\frac{5}{16}$	$3\frac{3}{8}$	$3\frac{5}{8}$	$3\frac{7}{8}$
29. ORDER	$3\frac{3}{4}$	4	$4\frac{3}{4}$	$3\frac{7}{8}$	$3\frac{11}{16}$	$4\frac{3}{8}$	$4\frac{1}{8}$	$3\frac{5}{8}$	$3\frac{7}{8}$	1	$4\frac{5}{16}$
30. EVENT.	$4\frac{1}{16}$	$4\frac{3}{8}$	4	$4\frac{1}{4}$	$4\frac{1}{16}$	$4\frac{3}{8}$	$4\frac{5}{16}$	$3\frac{13}{16}$	$4\frac{5}{16}$	$4\frac{1}{8}$	$4\frac{9}{16}$
32. TUNE	$3\frac{9}{16}$	$3\frac{3}{4}$					$3\frac{7}{8}$	$3\frac{7}{16}$	$3\frac{3}{8}$	$3\frac{3}{4}$	4
34. COMP.	$4\frac{3}{16}$	$4\frac{3}{4}$	$4\frac{3}{8}$	$4\frac{7}{16}$	$4\frac{5}{8}$	$4\frac{5}{8}$	$4\frac{1}{2}$	$4\frac{1}{4}$	$4\frac{1}{2}$	$4\frac{9}{16}$	$4\frac{11}{16}$
35. CAUS.	$4\frac{3}{16}$	$4\frac{3}{4}$	$4\frac{5}{16}$	$4\frac{3}{8}$	$4\frac{9}{16}$	$4\frac{9}{16}$	$4\frac{1}{2}$	$4\frac{1}{4}$	$4\frac{7}{16}$	$4\frac{9}{16}$	$4\frac{11}{16}$

Admeasurements of Skulls.	Chinese.	Karanqua Chief.*	Seminole Chief.*	Indian Squaw.*	Indian Princess.*	Indian Chief.*	Hottentot Fem.*	African slave.*	Chinook Indian, Columbia river	Chinook Indian, Columbia river	Idiot Female.*
1. AMAT.	$3\frac{1}{2}$	$2\frac{5}{8}$	$2\frac{3}{8}$	$2\frac{3}{8}$	$3\frac{1}{4}$	$3\frac{5}{8}$	$3\frac{1}{2}$	$2\frac{3}{4}$	$3\frac{1}{4}$		3
2. PHILO.	4	$3\frac{5}{8}$	$3\frac{1}{8}$	$3\frac{1}{4}$	$3\frac{3}{4}$	$3\frac{7}{8}$	$4\frac{1}{4}$	$3\frac{3}{4}$	$4\frac{5}{8}$	$3\frac{5}{8}$	$3\frac{3}{4}$
3. CONCE.	$4\frac{3}{8}$	$4\frac{9}{16}$	4	$3\frac{5}{8}$	$4\frac{7}{16}$	$4\frac{3}{8}$	$4\frac{1}{4}$	$4\frac{1}{8}$	$4\frac{1}{4}$	$4\frac{1}{16}$	4
4. ADHES.	$3\frac{5}{8}$	$3\frac{1}{2}$	$3\frac{1}{4}$	$3\frac{1}{8}$	$3\frac{3}{4}$	$3\frac{5}{8}$	4	$3\frac{1}{2}$	$4\frac{1}{2}$	$3\frac{1}{2}$	$3\frac{1}{2}$
5. COMBA.	$2\frac{1}{2}$	3	$2\frac{7}{8}$	$2\frac{7}{8}$	3	3	$2\frac{7}{8}$	$2\frac{5}{8}$	3	$3\frac{1}{8}$	$2\frac{1}{4}$
6. DESTR.	$2\frac{5}{16}$	$3\frac{5}{8}$	$3\frac{1}{2}$	$2\frac{1}{2}$	$2\frac{5}{8}$	$2\frac{11}{16}$	$2\frac{3}{8}$	$2\frac{1}{4}$	$3\frac{3}{4}$	$2\frac{7}{8}$	$1\frac{3}{4}$
ALIM.	$2\frac{1}{4}$								$2\frac{5}{8}$	3	
7. SECRE.	3	3	$2\frac{7}{8}$	$3\frac{1}{8}$	$3\frac{1}{8}$	$3\frac{3}{8}$	$2\frac{3}{4}$	$2\frac{3}{4}$	$3\frac{1}{4}$	$3\frac{1}{2}$	2
8. ACQUI.	$3\frac{1}{4}$	$3\frac{1}{2}$	$3\frac{3}{8}$	$3\frac{7}{16}$	$3\frac{1}{2}$	$3\frac{1}{2}$	$3\frac{1}{8}$	$3\frac{1}{8}$	$3\frac{1}{2}$	$3\frac{5}{8}$	$2\frac{3}{8}$
9. CONST.	$3\frac{3}{8}$	$3\frac{3}{8}$	$3\frac{1}{8}$	$3\frac{5}{8}$	$3\frac{1}{2}$	$3\frac{3}{8}$	$3\frac{1}{4}$	$3\frac{1}{4}$	$3\frac{1}{2}$		$2\frac{1}{2}$
10. SELF-E.	$4\frac{11}{16}$	$4\frac{7}{8}$	$4\frac{1}{8}$	$4\frac{7}{16}$	5	5	$4\frac{3}{4}$	$4\frac{9}{16}$	$4\frac{5}{8}$	$4\frac{13}{16}$	4
11. APPRO.	$4\frac{1}{2}$	$4\frac{3}{4}$	$4\frac{1}{4}$	$4\frac{1}{2}$	$4\frac{3}{16}$	$4\frac{7}{8}$	$4\frac{1}{4}$	$4\frac{3}{4}$		$4\frac{5}{8}$	4
12. CAU.	$4\frac{1}{8}$	$4\frac{1}{4}$	4	$4\frac{3}{8}$	$4\frac{1}{4}$	$4\frac{9}{16}$	$4\frac{3}{8}$	$4\frac{1}{2}$	$4\frac{1}{2}$	$4\frac{3}{8}$	$3\frac{5}{8}$
13. BENEV.	$4\frac{1}{2}$	$4\frac{7}{8}$	$4\frac{3}{8}$	$5\frac{1}{8}$	$4\frac{7}{8}$	$5\frac{1}{16}$	$4\frac{5}{8}$	$4\frac{7}{16}$	$3\frac{3}{4}$	$4\frac{3}{4}$	4
14. VENER.	$4\frac{9}{16}$	5	$4\frac{3}{8}$	5	5	$5\frac{1}{8}$	$4\frac{5}{8}$	$4\frac{7}{16}$	4	$4\frac{1}{16}$	$4\frac{1}{8}$
15. FIRM.	$4\frac{5}{8}$	$5\frac{1}{16}$	$4\frac{1}{2}$	$4\frac{3}{4}$	$5\frac{1}{16}$	$5\frac{3}{16}$	$4\frac{7}{8}$	$4\frac{1}{2}$	$4\frac{5}{16}$	$4\frac{7}{8}$	$4\frac{1}{4}$
16. CONSCI.		$4\frac{7}{8}$	$4\frac{1}{2}$	$4\frac{13}{16}$	$4\frac{15}{16}$	$5\frac{1}{16}$	$4\frac{7}{8}$	$4\frac{5}{8}$		$4\frac{1}{16}$	$4\frac{3}{16}$
17. HOPE	$4\frac{1}{2}$	$4\frac{7}{8}$	$4\frac{7}{16}$	$4\frac{1}{16}$	$4\frac{15}{16}$	$5\frac{1}{16}$	$4\frac{5}{8}$	$4\frac{7}{16}$	$4\frac{1}{8}$	$4\frac{3}{4}$	$4\frac{1}{8}$
18. MARV.									$3\frac{13}{16}$	$4\frac{3}{8}$	
19. IDEAL.	$4\frac{1}{4}$	$4\frac{7}{16}$	$4\frac{1}{8}$	$4\frac{5}{8}$	$4\frac{3}{8}$	$5\frac{5}{8}$	$4\frac{3}{8}$	$4\frac{1}{2}$	4	$4\frac{3}{8}$	$3\frac{1}{8}$
20. MIRTH	$4\frac{1}{4}$	$4\frac{3}{8}$	$4\frac{1}{8}$	$4\frac{11}{16}$	$4\frac{3}{8}$	$4\frac{5}{8}$	$4\frac{3}{8}$	4	$3\frac{3}{4}$	$4\frac{9}{16}$	$3\frac{5}{8}$
21. IMITA.	$4\frac{7}{16}$	$4\frac{3}{4}$	$4\frac{3}{8}$	5	$4\frac{1}{16}$	$4\frac{7}{8}$	$4\frac{4}{8}$	$4\frac{7}{16}$	$3\frac{3}{4}$	$4\frac{6}{8}$	4
22. INDIV.	$3\frac{13}{16}$	$4\frac{1}{8}$	$3\frac{3}{4}$	$4\frac{7}{16}$	$4\frac{1}{8}$	$4\frac{1}{8}$	$4\frac{1}{8}$	$3\frac{1}{2}$	$3\frac{1}{16}$	$4\frac{3}{8}$	$3\frac{1}{8}$
27. LOCAL.	$3\frac{7}{8}$	$4\frac{1}{4}$	4	$4\frac{7}{16}$	$4\frac{1}{4}$	$4\frac{5}{16}$	$4\frac{5}{8}$	3	$3\frac{9}{16}$	$4\frac{3}{8}$	$3\frac{3}{8}$
28. CALCU.	$3\frac{1}{2}$	$3\frac{3}{4}$	$3\frac{5}{8}$	$3\frac{7}{8}$	$3\frac{5}{8}$	$3\frac{5}{8}$	$4\frac{3}{8}$	3	4	$3\frac{3}{4}$	3
29. ORDER	$3\frac{5}{8}$	4	3		$3\frac{7}{8}$	$3\frac{7}{8}$	4	$3\frac{3}{8}$	$4\frac{5}{8}$	$3\frac{15}{16}$	$3\frac{1}{4}$
30. EVENT.	$3\frac{7}{8}$	$4\frac{1}{4}$				$4\frac{1}{2}$	$4\frac{3}{8}$	$3\frac{11}{16}$		$4\frac{1}{2}$	$3\frac{3}{8}$
32. TUNE	$3\frac{3}{8}$								$3\frac{3}{8}$	$3\frac{3}{4}$	
34. COMP.	$4\frac{1}{4}$							$4\frac{1}{16}$	$3\frac{5}{8}$	$4\frac{5}{8}$	$3\frac{3}{4}$
35. CAUS.	$4\frac{1}{4}$							$4\frac{1}{16}$		$4\frac{9}{16}$	$3\frac{1}{16}$

Admeasurements of Heads of Indian Chiefs of N. W. Territory.	Keo-Kuck, Chief of Sacs & Foxes.	Aquassau, wife of Fox-head Brave	Daughter of the Old She Bear.	Le Clare, ¼ blood U. S. Interpreter.	Wappanousa, or Child Chief.	Pashepaho, the Stabber.	Keonam, Talker for Nothing.	Keckkonwinney, or the Elk Horns.	Nanankee, or the Reformer.	Wapashancan, or White Skin.	Mechasisqua.
1. AMAT.	3 1/8	2 1/2	2 1/8	3 5/8	3 1/8	3 1/8	2 3/4	3	3 3/8	3 1/2	2 3/4
2. PHILO.	3 7/8	3 5/8	3 1/4	4 1/4	3 3/8	3 1/2	3 1/4	3 3/4	4 1/8	4 1/8	3 3/4
3. CONCE.	4 1/2	4 1/2	4 1/4	5 1/8	4 1/2	4 1/2	4 7/8	5	4 5/8	5	4 3/4
4. ADHES.	4	3 3/4	3 1/4	4 3/8	3 3/4	4	4 1/8	4 1/4	4 1/4	4 3/8	3 7/8
5. COMBA.	3 5/8	3 1/4	3	3 5/8	3 3/8	3 3/8	3 3/8	3 5/8	3 5/8	3 5/8	3
6. DESTR.	3 1/2	3 1/4	2 3/4	3 1/2	3 5/8	3 5/8	3 1/4	3 3/8	3 5/8	3 3/8	3
ALIM.											
7. SECRE.	4 1/4	3 3/4	3 1/4	3 3/4	4	4 1/8	4	4 1/8	4 1/8	4 1/4	3 5/8
8. ACQUI.	4 1/4	3 1/2	3 1/16	4	4 1/2	4 3/8	3 7/8	4 3/4	4	4 3/8	3 5/8
9. CONST.	4	3 1/2	3 1/8	3 3/4	4 1/8	4	3 3/4	3 3/8	3 3/4	4 3/8	3 3/8
10. SELF-E.	5 3/8	5 1/4	5 1/8	5 1/2	5 1/4	5 3/8	5 1/4	5 5/8	5 3/8	5 1/2	5 1/4
11. APPRO.	5 1/4	5 1/4	5	5 3/8	5	5 3/8	5	5 1/2	5 3/8	5 1/2	5
12. CAU.	5 1/8	5	4 5/8	5 1/8	4 7/8	5	4 3/[illegible]	5 1/4	5	5	4 3/8
13. BENEV.	5 5/8	5 1/8	4 1/8	5 3/8	5 3/8	5 5/8	[illegible] 3/8	5 3/16	5 3/8	5 5/8	5 3/8
14. VENER.	5 3/4	5 3/8	5 1/8	5 5/8	5 1/2	5 3/4	5 1/2	5 1/8	5 1/2	5 3/4	5 3/8
15. FIRM.	5 3/4	5 1/2	5 1/4	5 5/8	5 1/2	5 7/8	5 5/8	5 7/8	5 5/8	6	5 5/8
16. CONSCI.	5 5/8	5 3/8	5 1/16	5 1/2	5 3/8	5 5/8	5 1/4	5 5/8	5 3/8	5 5/8	5 1/4
17. HOPE	5 5/8	5 1/8	5	5 3/8	5 1/4	5 1/2	5 3/8	5 3/4	5 3/8	5 5/8	5 1/8
18. MARV.											
19. IDEAL.	5 1/8	4 3/4	4 3/8	4 3/4	4 3/8	4 3/4	4 3/4	4 3/4	4 3/4	5	4 3/8
20. MIRTH	4 7/8	4 1/4	3 7/8	4 3/4	4 5/8	5	4 7/8	5	4 5/8	5 3/8	4 3/8
21. IMITA.	5 1/2	5 1/8	4 5/8	5 1/8	5 3/16	5 3/8	5 3/16	5 1/4	5 1/4	5 3/8	5 1/8
22. INDIV.	4 3/4	3 7/8	3 3/4	4 3/8	4 3/4	4 3/4	4 3/4	4 3/4	4 1/2	4 5/8	4 3/4
27. LOCAL.	4 7/8	4 3/8	3 1/8	4 1/2	4 7/8	5	4 7/8	4 7/8	4 5/8	4 5/8	4 1/2
28. CALCU.	4 1/8	4	3 1/4	3 3/4	4 1/16	4	4 1/8	4 1/8	4	4 1/8	3 3/4
29. ORDER	4 3/8										
30. EVENT.	5	4 1/2	4	4 1/2	5	5	4 7/8	5	4 3/4	4 7/8	
32 TUNE	4	3 3/4	3 3/8	4	4	4 1/8	4 1/4	4 1/8	4	4	3 5/8
34. COMP.	5 1/4	4 7/8	4 1/2	5	5 1/4	5 1/4	5 1/4	5 3/8	5 1/8	5 1/4	4 7/8
35. CAUS.	5 1/8	4 3/4	4 3/8	4 7/8	5 1/16	5 1/8	5	5 1/4	5	5 1/8	4 1/8

Admeasurements of Heads and Skulls.	G. Combe's approved bust.	F. Coombs, Phrenologist.	Ancient Peruvian, from Arica.	Ancient Peruvian, from Arica	Ancient Peruvian, from Arica	Ancient Peruvian Child.	American Idiot.	London Idiot.	Orang.	Baboon.
1. AMAT.	$3\frac{1}{4}$	$3\frac{1}{8}$	$3\frac{1}{2}$	$3\frac{3}{8}$	$3\frac{1}{8}$	3	$2\frac{3}{4}$	$2\frac{6}{16}$	2	1
2. PHILO.	4	$3\frac{7}{8}$	$4\frac{5}{8}$	$4\frac{1}{2}$	4	$4\frac{1}{2}$	$3\frac{1}{4}$	$3\frac{5}{16}$	$2\frac{5}{8}$	$1\frac{1}{4}$
3. CONCE.	$4\frac{1}{2}$	$4\frac{3}{8}$	5	$4\frac{12}{4}$	$4\frac{1}{2}$	5	$3\frac{1}{2}$	$3\frac{4}{8}$	$2\frac{3}{4}$	$1\frac{3}{16}$
4. ADHES.	$4\frac{1}{16}$	4	$4\frac{1}{4}$	$4\frac{1}{2}$	$3\frac{7}{8}$	$4\frac{1}{4}$	$3\frac{3}{16}$	$3\frac{3}{8}$	$2\frac{3}{8}$	$1\frac{1}{4}$
5. COMBA.	$3\frac{1}{4}$	$3\frac{3}{8}$	$3\frac{5}{16}$	$2\frac{7}{8}$	$2\frac{7}{8}$	$2\frac{3}{8}$	$2\frac{3}{8}$	$2\frac{6}{8}$	2	$1\frac{1}{8}$
6. DESTR.	$3\frac{3}{16}$	3	$2\frac{3}{8}$	$2\frac{5}{16}$	$2\frac{1}{4}$	$1\frac{3}{4}$	$2\frac{1}{4}$	$2\frac{5}{8}$	$1\frac{11}{16}$	$1\frac{1}{16}$
ALIM.		3	$2\frac{1}{4}$	$2\frac{1}{8}$	$2\frac{1}{4}$					
7. SECRE.	$3\frac{3}{4}$	$3\frac{5}{8}$	$3\frac{1}{8}$	$2\frac{7}{8}$	$2\frac{7}{8}$	$2\frac{3}{8}$	$2\frac{5}{8}$	$2\frac{7}{8}$	$1\frac{3}{4}$	$1\frac{3}{16}$
8. ACQUI.	$3\frac{1}{16}$	$3\frac{7}{8}$	3		3	$2\frac{5}{8}$	$2\frac{5}{8}$	$2\frac{7}{8}$	$2\frac{1}{2}$	$1\frac{1}{4}$
9. CONST.	$3\frac{9}{16}$	4	$3\frac{3}{8}$		$3\frac{1}{16}$		$2\frac{1}{8}$	3	$2\frac{3}{8}$	$1\frac{3}{8}$
10. SELF-E.	$5\frac{1}{8}$	$5\frac{1}{8}$	$5\frac{3}{16}$	$5\frac{1}{4}$	$5\frac{1}{8}$	5	$3\frac{11}{8}$	$3\frac{5}{8}$	3	$1\frac{3}{8}$
11. APPRO.	$5\frac{1}{16}$	5		5	$5\frac{1}{16}$	$4\frac{7}{8}$	$3\frac{5}{8}$	$3\frac{1}{2}$	$2\frac{7}{8}$	$1\frac{3}{8}$
12. CAU.	$4\frac{5}{8}$	$4\frac{9}{16}$	$4\frac{3}{4}$	$4\frac{1}{2}$	$4\frac{1}{2}$	$4\frac{1}{2}$	$3\frac{1}{8}$	$3\frac{3}{16}$	$2\frac{3}{4}$	$1\frac{3}{16}$
13. BENEV.	$5\frac{1}{8}$	$5\frac{5}{8}$	$4\frac{7}{8}$	$4\frac{3}{8}$	$4\frac{11}{16}$	4	$3\frac{1}{4}$	$3\frac{3}{8}$	3	$1\frac{7}{8}$
14. VENER.	$5\frac{1}{8}$	$5\frac{9}{16}$	5	$4\frac{9}{16}$	$4\frac{13}{16}$	$4\frac{1}{4}$	$3\frac{5}{16}$	$3\frac{1}{2}$	3	$1\frac{3}{4}$
15. FIRM.	$5\frac{1}{4}$	$5\frac{9}{16}$	$5\frac{3}{16}$	5	$5\frac{1}{16}$	$4\frac{3}{4}$	$3\frac{9}{16}$	$3\frac{9}{16}$	3	$1\frac{5}{8}$
16. CONSCI.	$5\frac{1}{16}$	$5\frac{5}{16}$	$5\frac{1}{8}$	$4\frac{7}{8}$	5	$4\frac{11}{16}$	$3\frac{5}{8}$	$3\frac{3}{16}$	$2\frac{15}{8}$	$1\frac{9}{16}$
17. HOPE	$4\frac{5}{16}$	$5\frac{1}{4}$	5	$4\frac{1}{2}$	$4\frac{1}{16}$	$4\frac{3}{8}$	$3\frac{1}{2}$	$3\frac{1}{2}$	$2\frac{15}{16}$	
18 MARV.		$5\frac{1}{16}$	$4\frac{13}{16}$	$4\frac{7}{16}$	$4\frac{1}{2}$	4				
19 IDEAL.	$4\frac{3}{8}$	$4\frac{1}{8}$	$4\frac{1}{2}$	4	$4\frac{3}{16}$	$3\frac{5}{8}$	$3\frac{1}{4}$	$3\frac{3}{16}$	$2\frac{3}{4}$	$1\frac{1}{8}$
20 MIRTH	$4\frac{3}{8}$	$4\frac{1}{4}$	$4\frac{9}{16}$	$3\frac{3}{4}$	$4\frac{3}{16}$	$3\frac{5}{8}$	3	$3\frac{3}{8}$	$2\frac{7}{8}$	$1\frac{1}{8}$
21 IMITA.	5	$5\frac{1}{4}$	$4\frac{5}{8}$	4	$4\frac{1}{2}$	$4\frac{1}{16}$	$3\frac{1}{8}$	$3\frac{5}{16}$	$2\frac{1}{8}$	$1\frac{3}{4}$
22. INDIV.	$4\frac{1}{16}$	$4\frac{5}{16}$	4	$3\frac{9}{16}$	$3\frac{1}{16}$	$2\frac{7}{8}$	$2\frac{7}{8}$	$3\frac{9}{16}$	$2\frac{7}{8}$	$2\frac{3}{8}$
27. LOCAL.	$4\frac{3}{16}$	$4\frac{9}{16}$	4	$3\frac{9}{16}$	$3\frac{1}{8}$	3	$2\frac{7}{8}$	$3\frac{9}{16}$	$2\frac{7}{8}$	$2\frac{3}{8}$
28. CALCU.	$3\frac{3}{16}$	4	$3\frac{1}{2}$	$3\frac{1}{2}$	$3\frac{3}{8}$	$2\frac{1}{4}$	$2\frac{5}{8}$	$3\frac{5}{16}$	$2\frac{1}{2}$	
29. ORDER	$3\frac{13}{16}$	$4\frac{5}{16}$	$3\frac{13}{16}$	$3\frac{5}{8}$	$3\frac{9}{16}$	$2\frac{3}{4}$	$2\frac{1}{4}$	$3\frac{3}{4}$	$2\frac{3}{4}$	
30. EVENT.	$4\frac{3}{8}$	$4\frac{5}{8}$	$4\frac{1}{16}$	$3\frac{5}{8}$	$3\frac{3}{4}$	3	$2\frac{15}{16}$	$3\frac{3}{8}$	$2\frac{15}{16}$	
32. TUNE	$3\frac{13}{16}$	$4\frac{1}{4}$	$3\frac{5}{8}$	$3\frac{1}{16}$	$3\frac{5}{16}$	$2\frac{3}{8}$				
34. COMP.	$4\frac{3}{16}$	$5\frac{1}{8}$	$4\frac{1}{4}$	$3\frac{3}{4}$	$3\frac{7}{8}$	$3\frac{1}{4}$	$3\frac{1}{16}$	$3\frac{5}{16}$	3	$2\frac{1}{16}$
35. CAUS.	$4\frac{1}{16}$	$5\frac{1}{16}$	$4\frac{1}{4}$	$3\frac{3}{4}$	$3\frac{13}{16}$	$3\frac{1}{4}$	3	$3\frac{1}{4}$	$2\frac{5}{16}$	$2\frac{1}{16}$

Admeasurements of Heads.	Cast of Robinson, the Murderer.	3 3/4 exc. of animal in McL's head.	Alex'r McLeod, the Loyalist.	W. L. Mackenzie, the Patriot.	13 1/2 excess of intellect in Mackenzie's head.	Murderer, Lieutenant R.	Dr. J. A. Houston, a good Reformer.	Mr. Thom, an honest Chartist.	E. A. Theller, a good Patriot.	Rev. Dr. Channing, of Boston.	B. Caunt, Champion of England.
1. Amat.	3 3/4	1	3 1/8	3 1/8		3 5/8	2 3/4	3 1/16	3 5/8	3	3 3/8
2. Philo.	4 1/8		4 3/8	4 3/4	3/4	4 1/4	4 1/2	4 3/8	4 3/8	4	4 1/2
3. Conce.	4 3/4		4 3/4	5 1/8	3/4	5 1/4	5 1/16	5 5/8	4 7/8	4 1/8	5 1/4
4. Adhes.	4 1/8	1/4	4 1/4	4 1/8		4 1/2	4	3 5/8	4 1/16	3 7/8	4 3/8
5. Comb.	3 1/2	1	3 5/8	3 1/8		3 7/8	3	3 1/8	3 3/4	3 3/8	3 5/8
6. Destr.	3 1/4	1	3 9/16	3 1/16		3 5/8	3	3	3 5/16	3 1/16	3 1/8
Alim.	3 1/4	1/2	3 1/8	3 1/8			3	3	3 5/16		3 1/4
7. Secre.	3 3/4	3/4	4	3 5/8		4 1/8	3 5/8	3 5/8	4	3 3/4	4
8. Acqui.	3 7/8		3 5/8	3 5/8		4 1/8	3 1/2	3 1/2	4	4	4 1/4
9. Const.	3 3/4		3 3/4	3 3/4		4 3/8	3 5/8	4 1/16	4	3 7/8	4 1/4
10. Self-E.	5 1/8		5 1/4	5 5/8	3/4	5 1/2	5 5/8	5 1/2	5 3/16	5	5 5/8
11. Appro.	5		5 1/8	5 1/2	3/4	5 3/8	5 3/8	5 1/4	5 3/16	5 1/16	5 1/4
12. Cautio.	5		4 1/2	4 3/4	1/2	4 7/8	4 11/16	4 7/8	4 13/16	4 5/8	4 3/4
13. Benev.	5		5 1/8	5 1/2	3/4	5 3/8	5 1/2	5 13/16	5 1/8	5 3/8	5 1/2
14. Vener.	5 1/16		5 1/4	5 5/8	3/4	5 3/8	5 11/16	5 7/8	5 3/8	5 3/8	5 5/8
15. Firm.	5 1/8		5 1/2	5 3/4	1/2	5 7/16	5 7/8	6	5 1/2	5 1/2	6
16. Consci.	5 1/16		5 5/16	5 5/8	5/8	5 5/16	5 3/4	5 13/16	5 3/8	5 1/4	5 3/4
17. Hope.	4 7/8		5	5 1/2	1	5 1/4		5 5/8	5 5/16	5 1/4	5 5/8
18. Marv.	4 7/8		4 7/8	5 3/8	1		5 3/8	5 5/8	5 1/4	5 1/8	5 3/8
19. Ideal.	4 5/8		4 1/2	5	1	5 1/8	4 7/8	5	4 1/2	4 5/8	4 7/8
20. Mirth.	4 3/4		4 1/2	5	1	5 3/16	4 5/8	5 3/4	4 5/8	4 5/8	5
21. Imita.	4 7/8		4 3/4	5 1/4	1	5 1/8	5 5/16	5 5/8	5 1/4	5 1/8	5 3/8
22. Indiv.	4 3/8		4 1/8	4 1/4	1/4	4 3/4	4 9/16	4 1/16	4 1/8	4 1/4	4 5/8
27. Local.	4 1/8		4 3/8	4 1/2	1/4	5 1/8	4 11/16	4 7/8	4 1/2	4 3/8	4 7/8
28. Calcu.	3 3/4	1/2	4	3 3/4		4 1/8	4 1/16	3 7/8	3 3/4	3 3/4	4
29. Order.	4 3/8	1/4	4 1/8	4			4 5/16	4 3/8	4	4	4 1/2
30. Event.	4 9/16		4 3/8	4 3/8		5 1/8	4 11/16		4 1/2	4 5/8	4 7/8
32. Tune.	4 9/16			4 1/2		4 3/8			4 5/16		
34. Comp.	4 13/16		4 5/8	5 1/8	1	5 3/8	5 1/4	5 7/16	4 7/8	5	5 3/16
35. Caus.	4 3/4		4 5/8	5 1/16	7/8	5 1/4	5 1/8	5 5/16	4 13/16	5	5 1/8

The Bilio.
The Nerv.
The Sang.
The Lymp.

On comparing the measurements of Heads in this Table, it will be found, the Head of Alexander McLeod bears a more striking resemblance to that of Robinson the murderer, than even to that of Lieut Richards the murderer; the latter having more intellect than Robinson or McLeod.

Explanation of the Mode of Reading the Character.

Each of the thirty-five organs are described in the degrees of very large, large, full, moderate, and small, or very small. The *written* figures placed in the margin indicate the respective sizes of each organ in the individual examined— 6, or 5½ being the maximum, and 2, or 1 the minimum. For instance, if an organ is marked VERY LARGE, it will be indicated by the figures 6 or 5½ being placed in the margin opposite very large, 6 expressing a more active degree than 5½, and when marked so high, it is somewhat liable to abuse, or excessive action, unless the antagonist organs are also very large. When the organ or faculty is marked 5, or 4½, it is then LARGE, and presents a strong feature of character, but not to abuse or inordinate activity, unless the intellect or sentiments are very deficient. When numbered 4, or 3½ the faculty is then FULL, but does not present great activity, and may be considered above mediocrity. If marked 3, or 2½, only, it is then but MODERATE, or rather inactive, and implies a deficiency of the organ, and not easily roused to action. If marked as low as 2, or 1, it is then minimum, SMALL, or VERY SMALL, and infers an extremely inferior or deficient development in organs so designated.

To assist the student, the dot is placed opposite the particular line applicable to the character of the person examined, which line he is to read as his own character. The same scale of figures is also applied to the temperaments, and the average size of head and degree of activity.

The organ of Tune, and some few others, the author does not usually mark, as the temporal muscle and ridge present an impossibility of ascertaining exactly the size of these organs.

In estimating the phrenological character, it may be found that many of the organs are marked very large, from which we have to fear an excessive action. In such case it is desirable that a check should be placed on the activity of such organ or organs, (which, indeed, is the great use of a phrenological examination,) by the continual exercise of those of a dissimilar character. On the contrary, those which are marked small, or moderate, require stimulating as far as practicable. Those organs which require repressing in their action will be marked with a *dash* ——— below the figure, and those which need stimulating will have a *dot* · above the figure. The delineation of character by this mode may also sometimes appear contradictory: thus, when the organs of Benevolence and Acquisitiveness are both marked very large, the description would then seem to imply that such a person would be both miserly and generous. Such betray often a penny-wise and pound-foolish disposition. The same may also be said of Destructiveness, or passion and severity coupled with great good nature. Such an one will be as free to forgive as to resent an injury. These feelings are by no means incompatible in the same person.

The author has designated by marks several portions of the brain, the functions of which are not fully ascertained. From recent observations he is inclined to the opinion there is a semi-intellectual function or organ, Suavity, (*) between 13 and 34. The mark (?) under 17, the organ of Sublimity. Under 3, the mark (‡) is by some marked Inhabitiveness, which he does not mark. The star (*) behind the ear, is Vitativeness, (doubtful.) The (*) before it, is Alimentiveness, confirmed by all phrenologists. The star (*) beneath the eyes, one phrenologist only conceives to be the organ of Chemicality. (Improbable.)

The organ of Tune the author declines marking, unless it is extraordinary small or large, in consequence of the impossibility of ascertaining its exact size, from the thickness of the temporal muscle which covers it. The temporal ridge, or nickening of the skull in adult age, also presents a difficulty

PHRENOLOGICAL AND PHYSIOLOGICAL ANALYSIS OF CHARACTER, BY

CLASSIFICATION OF THE ORGANS.

ORDER 1—Feelings. GENUS 1—Propensities.
1.—AMATIVENESS.

Very Large—An extreme fondness; blind, passionate admiration.
Large—Very susceptible to the passion of love; strong affection.
Full—Very apt to become enamored, but inclined to be constant.
Moderate—Rather fastidious in selecting a lover; cold and reserved.
Small—Old maidenish; extremely particular; want of feeling.

Amativeness very large, combined with very large intellect as in profile No. 1, contrasted with Idiot, No. 2, having very small Amativeness and less Intellect.

Physiognomical language—the whole head and person, particularly the eyes, are intensely riveted and directed towards the object desired.

Uses—tenderness, kindness, and sympathy for the opposite sex, propagation of the species.

Abuses—immoderate, blind passion, no discrimination, for which reason Love is painted blind.

Location—in the posterior base of the head, just above where the hair terminates. The size may be ascertained by the thickness of neck and head behind the ears; usually much the largest in males.

> "Things base and vile, holding no quality,
> Love can transpose to form and dignity.
>
> Love looks not with the eyes, but with the mind—
> He says he loves my daughter;
> I think so too, for never gazed the moon
> Upon the water, as he 'll stand and read
> As 't were my daughter's eyes; and, to be 'lain,
> I think there is not half a kiss to choose
> Who loves another best."

No. 1. A large, fairly balanced Head. Large in the Intellect and very large Amativeness

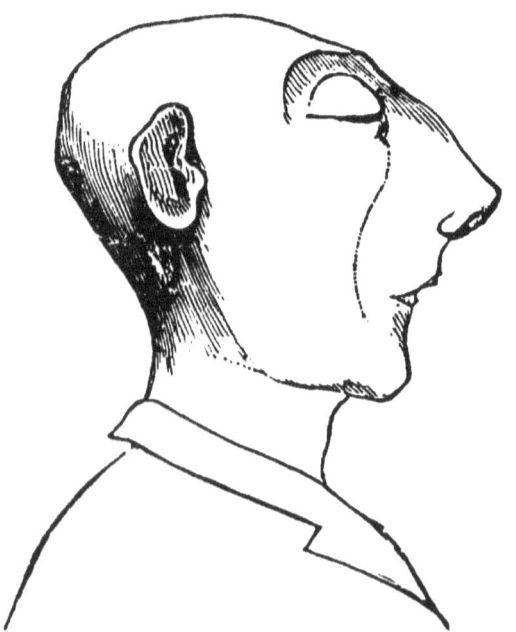

No. 2. Idiot, with small brain, but very large face.

25

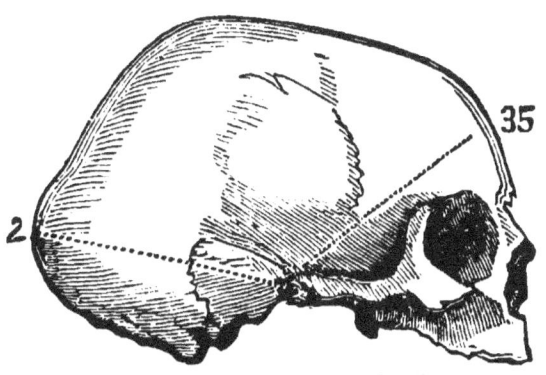

No. 3. Philoprogenitiveness, or maternal love, only full.

No 4. Skull of a Negress excessively fond of children. Exceedingly large in Philoprogenitiveness, or No. 2.

2.—PHILOPROGENITIVENESS.

Very Large—Excessive fondness for children; too indulgent.
Large—Strong degree of parental affection and tenderness.
Full—A due regard for children, but not a blind partiality.
Moderate—Indifference to children and pets; no anxiety for them
Small—Decided aversion to children; want of parental feeling.

Philoprogenitiveness full in lady with the babe, (No. 3,) and very large in skull of the negress, (No. 4,) who was only remarkable for spoiling children, in other respects nearly idiotic.

Physiognomical language—to incline the head towards the babe, and with a sweet smile caress it.

Uses—affectionate solicitude, providing for young.

Abuses—too indulgent, spoiling children, &c.

Location in the centre of back-head, a little above Amativeness; easily seen in females, in whom it is generally much larger than in males. Usually large in ladies who complain of difficulty in attaching their comb.

MOTHER'S LOSS OF HER CHILD.

"For, since the birth of Cain the first male child,
To him that did but yesterday suspire,
There was not such a gracious creature born.

—Grief fills the room up of my absent child,
Lies in his bed, walks up and down with me,
Puts on his pretty looks, repeats his words,
Remembers me of all his gracious parts,
Stuffs out his vacant garments with his form:
Then have I reason to be fond of grief."

3.—CONCENTRATIVENESS.

Very Large—Great power of riveting the attention, tedious, verbose.
Large—A talent for pursuing abstract or metaphysical questions.
Full—Ability to dwell on a subject and control the imagination.
Moderate—Versatility of thought and action; fond of variety.
Small—Inability to confine the attention, which is ever roving.

Concentrativeness and Inhabitiveness large, with very large Intellect and Ideality, in Milton, (No. 5,) who has produced one of the finest poems extant. He also severely lashed the vices of the age in which he lived, proving no less a patriot than a poet. Skull of savage, (No. 6,) small in each organ.

Physiognomical expression—the attention intently fixed, and the head bent forward.

Uses—gives power and continuity of emotions and ideas.

Abuses—tedious prolixity, and morbid dwelling on impressions, to the exclusion of external objects.

Location—above Inhabitiveness and below Self-Esteem, each side, on the longitudinal suture. Large in men of literary acquirements.

‡ INHABITIVENESS.

"Here woman reigns, the mother, daughter, wife,
Strews with fresh flowers the narrow vale of life;
In the clear heaven of her delighted eye,
An angel-guard of loves and graces lie;
Around her knees domestic duties meet,
And fireside pleasures gambol at her feet.
Where shall that land, that spot of earth be found?
Art thou a man?—a patriot?—look around;
Oh! thou shalt find, where'er thy footsteps roam,
This land thy country, and this spot thy home."

4.—ADHESIVENESS.

Very Large—Passionate and devoted in attachments to friends.
Large—Unalterable affection; enduring all things for love.
Full—Constancy; pure affection; platonic and sincere attachments
Moderate—Changeable in love or affection; very fond of variety.
Small—Destitute of pure affection; always desirous of change.

Adhesiveness, remarkably large in one of nature's fairest flowers, (No. 7,)—a pattern of constancy, affection, and sincere friendship.

Physiognomical expression—gently to incline the head towards the object of attachment, as in profile.

Uses—attachments, friendships, and love of family.

Abuses—too much kindness and affection even to worthless persons; constancy; true love.

Location—upward and outward from Philoprogenitiveness, generally very large in females.

"———Or bid me go into a new-made grave,
And hide me with a dead man in his shroud;
Things that, to hear them told, have made me tremble;
And I will do it without fear or doubt,
To live an unstain'd wife to my sweet love.

O! happy love! where love like this is found;
O! heartfelt rapture! bliss beyond compare!
I've paced much this weary, mortal round,
And sage experience bids me this declare—
"If heaven a draught of heavenly pleasure spare,"
One cordial in this melancholy vale,
'Tis when a youthful, loving, modest pair,
In other's arms breathe out the tender tale,
Beneath the milk-white thorn that scents the ev'ning gale."

No. 5.   Portrait of Milton the poet.

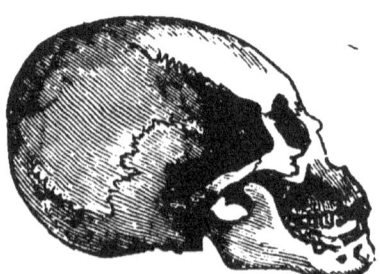

No. 6   Skull of a savage Hottentot. Very small in the Intellect and Sentiments.

No. 7. One of natures fairest flowers.

3*

5.—COMBATIVENESS.

Very Large—Ferocious courage, rage, anger, and violence.
Large—Great personal courage under opposition and danger.
Full—Resolution, decision, energy, and determination of charac er
Moderate—Indifferent courage; averse to extreme measures.
Small—Weak and irresolute; too much disposed to yield to others.

Combativeness very large in the Karanqua Texian Indian Chief, (No. 9,) with also very deficient intellect and sentiment, contrasted with a sketch of an author, (No. 10,) who above all things desires to cement the bonds of love, harmony, and friendship between the English and American people, who have but one common enemy, the British aristocracy, (the people are one.) National wars are wholesale murders, according to Phrenology and common sense.

Physiognomical expression—to carry the head menacingly; discordant, harsh voice, and features distorted when excited by rage.

Uses—Courage to resist and overcome difficulties.
Abuses—love of cruel and barbarous sports, &c.
Location—about one inch behind the top of the ear, giving great width around and behind the ears.

> "But when the blast of war blows in our ears,
> Then imitate the action of the tiger;
> Stiffen the sinews—summon up the blood—
> Disguise fair nature with hard-favored rage;
> Then lend the eye a terrible aspect—
> Let it pry through the portage of the head
> Like the brass cannon—let the brow o'erwhelm it,
> As fearfully as doth a galled rock
> O'erhang and jutty his confounded base,
> Swilled with the wild and wasteful ocean."

6.—DESTRUCTIVENESS.

Very Large—Cruel, ferocious, vindictive, revengeful, murderous.
Large—Passionate and hasty in expressions of anger; quarrelsome.
Full—Energetic, resolute, and decided; easily excited to action.
Moderate—Ability to control the passions; indolent and inactive.
Small—Extremely averse to action; lacks energy of character.

Destructiveness very large in the head of murderer, (pp. 9 10,) and deaf Burke; very small in the Hindoo skulls, (p. 10.) See also admeasurements.]

Physiognomical expression—when roused, furious gesticulation, brows contracted, the teeth set, the countenance distorted with passion.
Uses—desire to destroy for food, to kill noxious reptiles, &c.
Abuses—leads to passion, rage, severity and murder.
Location—around and above the top of the ears.

> "Would curses kill, as doth the mandrake's groan,
> I would invent as bitter-searching terms,
> As curst, as harsh, and horrible to bear,
> Deliver'd strongly through my fixed teeth,
> With full as many signs of deadly hate,
> As lean-faced envy in her loathsome cave;
> My tongue should stumble in my earnest words;
> Mine eyes should sparkle like the beaten flint;
> My hair be fixed on end, as one distract."

*ALIMENTIVENESS.

Very Large—An excessive fondness for high, luxurious living.
Large—Fond of banqueting; a hearty, keen, and healthy appetite.
Full—Good relish for food, but will not indulge in excesses.
Moderate—Abstemious; no desire for high-seasoned or dainty food.
Small—Disrelish for food, and prefers vegetable to animal diet.

Uses—appetite for food. Abuses—gluttony, intemperance

7.—SECRETIVENESS.

Very Large—Dissimulation; cunning; treacherous, lying, deceitful.
Large—Artful, designing, and intriguing; an adept at management.
Full—Prudent, cautious, and calculating; without much deception.
Moderate—Candid, open, communicative, agreeable, and sociable.
Small—Extremely liable to be imposed on; want of circumspection.

Secretiveness large in the murderer, and small in Hindoos. [See preceding cuts; also admeasurements.]
Physiognomical expression—cunning, stealthy, and gliding movements, like a cat; rolling the eyes without turning the head; downward look; slyness
Uses—to conceal emotions which involuntarily arise in the mind, until sanctioned by the judgment for utterance.
Abuses—slyness, cunning, duplicity, and lying.
Location—above, around, and adjoining the organ of Destructiveness

32

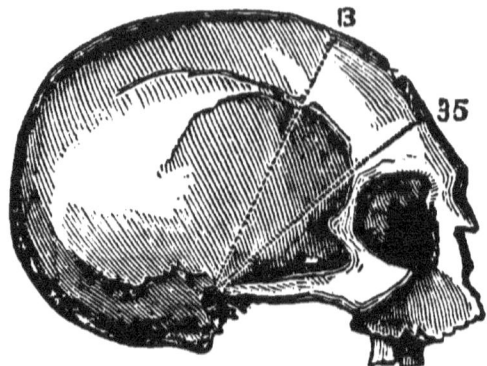

No. 9. Skull of a Cannibal Indian Chief of Texas. Small in 3 and 35

No. 10. Anti-Combativeness or the advocate of universal peace.

No. 13. Portrait of Peter the Great; a very singular King, having been a great benefactor to his subjects.

No. 14. Lord Byron; remarkable for his hatred of tyranny, as well as love for the beautiful, and poetical talents of the very highest order.

> "Why, I can smile, and murder while I smile,
> And cry content to that which grieves my heart,
> And wet my cheeks with artificial tears,
> And frame my face to all occasions;
> I'll drown more sailors thad the mermaid shall;
> I'll slay more gazers than the basilisk;
> I'll play the orator as well as Nestor;
> Deceive more slyly than Ulysses could,
> And like a Sinon, take another Troy."

8.—ACQUISITIVENESS.

Very Large—Extremely miserly, sordid, penurious, and thievish.
Large—Indefatigable in getting money, parsimonious, and saving.
Full—Industrious, frugal, and economical, occasionally liberal.
Moderate—Generous and free; little solicitude about acquiring.
Small—Indifferent about money, and very apt to spend it too freely

Acquisitiveness very large in misers, and generally in murderers.

Physiognomical expression—disagreeable contraction of the features; hands groping in the pockets; prying, peering inquisitiveness about dollars and cents; clutches things eagerly dreams about money.

Uses—to procure necessaries, articles of utility, &c.

Abuses—Miserly, avaricious, and grasping; theft, or swindling.

Location—forward of Secretiveness, and above Alimentiveness, giving width behind the temples.

> The grovelling wretch who barters souls for gold,
> Ne'er knew the blissful charities of generous love,
> His soul unlovely, and his body lean with anxious care,
> The widow's groan and orphan's tear he will not hear.
> Tottering with age, he still doth grasp for more,
> And like the yawning grave he ever hides his store;
> With fear and jealousy oppressed, he numbers o'er and o'er
> This shining dross, that drags his soul to endless woe.

9.—CONSTRUCTIVENESS.

Very Large—Great mechanical talents for building and inventing.
Large—Excellent judgment in planning and skill in contriving.
Full—Fair share of mechanical skill and ingenuity; good ability.
Moderate—Indifferent talents, and distaste for mechanical pursuits.
Small—Very bungling and awkward; great aversion to using tools.

Constructiveness very large in boy. [See form.] Also very large in Peter the Great, (No. 13,) who learnt and practised blacksmithing, toothdrawing, and nearly fifty other trades, to civilize his Russian subjects—greater than an Alexander, or a Cæsar. Very small in skull of savage.

Physiognomical language—passionately fond of handling and viewing things in mechanics; taste for building, fashioning by hand, &c. Prying and curious in workmanship.

Uses—to build houses, ships, and objects of utility.

Abuses—to make engines to destroy or injure others; fond of whittling door-posts, &c.

Location—this organ is easily seen when very large, by a great width in the temples.

THE FORMATION OF THE WORLD.

And in his hand
"He took the golden compasses, prepared
In God's eternal store, to circumscribe
This universe and all created things.
One foot he center'd, and the other turned
Round the vast profundity obscure,
And said, "Thus far extend, thus far thy bounds,
This be thy just circumference, O world."
Thus God the heaven created, thus the earth,
Matter unformed and void; darkness profound."

GENUS 2—Inferior Sentiments.

10.—SELF-ESTEEM.

Very Large—Presumptuous, proud, arrogant, and overbearing.
Large—Ambitious of distinction; independent, and high-minded.
Full—Proper degree of pride, and correct notions of propriety.
Moderate—Wanting in dignity and self-confidence; easily abashed
Small—Greatly underrates himself, and is too diffident.

Self-Esteem very large in Lord Byron, (No. 14,) combined with the highest Intellect and Ideality; also, in the female head on the following page, (15,) produced in a great measure by the flattery of the beaux. Oh, ladies, beware, and not wish to be a belle!

Physiognomical language—in man, a haughty and erect carriage, pride and disdain. Ladies are apt to toss the head, look disdainful, and so to spoil their pretty faces.

Uses—self-respect, independence, and dignity.

Abuses—pride, self-conceit, arrogant domineering.

Location—at the back of the upper part of the head or crown, where the hair turns.

No. 15. Self Esteem Illustrated.

No. 16. Smiling Approbativeness Illustrated.

No. 16½. Letitia Buonaparte, mother of Napoleon. Head and face, like Napoleon's, (page 74,) display masculine energy and determination of character in a remarkable degree.

> Disdain and scorn ride sparkling in her eyes,
> Misprising what they look on, and her wit
> Values itself so highly, that to her
> All matter else seems weak; she cannot love,
> Nor take no shape nor project of affection,"
> She is so self-endeared.

11.—APPROBATIVENESS.

Very Large—Too sensitive of the opinions of others; vain; showy.
Large—Ambitious of applause, and fond of being admired.
Full—A due regard for popularity, but not too subservient.
Moderate—Indifference to the observations of others; independent.
Small—Want of politeness and affability; careless and rude.

Approbation very large in the beautiful Miss (No. 16,) on preceding page, combined with every amiable sentiment, rendering her an universal favorite.

Physiognomical expression—extremely affable, pleasing, and sociable; the mouth is wreathed with an affectionate smile, and the eyes beaming with good nature; desire to oblige.

Uses—to please and conciliate; true, kind, and unsophisticated politeness; affability.

Abuses—vanity, ambition, rivalry, and most remarkably fond of flattery.

Location—on the top of head, on each side of Self-Esteem. Largest in females.

> "I have marked
> A thousand blushing apparitions start
> Into her face; a thousand innocent shames
> In angel whiteness bear away those blushes."

12.—CAUTIOUSNESS.

Very Large—Cowardice, fear, and timidity; cannot be made to fight.
Large—Indecision; want of resolution, courage, and determination.
Full—Prudent and cautious; enabled to act with great decision.
Moderate—Absence of fear, but rather improvident and careless
Small—Want of prudence; extremely rash, liable to accidents.

Caution very large in Hindoo, [see cut and admeasurements,] small in Pugilist, deaf Burke, and Murderer, (pp. 9,10.)

Physiognomical expression—restless, anxious eyes and careworn features; quiet, subdued manner, too diffident and timid.

Uses—to provide against danger, accidents and difficulties; to anticipate the future.

Abuses—inordinate fear, timidity, cowardice and irresolution.

Location—in the middle of the parietal bone, about three inches above the ears, and somewhat behind them.

"Well, 'tis no matter; honor pricks me on. Yea, but how if honor prick me off when I come on? how then, can honor set to a leg? No. Or an arm? No. Or take away the grief of a wound? No. Honor hath no skill in surgery then? No. What is honor? A word. What is in that word? Honor. What is that honor? A trim reckoning. Who hath it? he that died o'Wednesday, doth he feel it? No. Doth he hear it? No. Is it insensible then? Yea to the dead. But will it not live with the living? No. Why? Detraction will not suffer it, therefore I'll none of it: honor is a mere escutcheon. and so ends my catechism."

GENUS 3—Superior Sentiments.
13.—BENEVOLENCE.

Very Large—Munificent and generous to a fault; sympathetic.
Large—Free, kind, and liberal; tender, charitable, and humane.
Full—Active desire of doing good; great degree of sympathy.
Moderate—Indifferent to the welfare of others; selfish and unkind
Small—Sordid, avaricious, mean, and totally insensible to charity.

Benevolence very large in the amiable lady illustrated on next page a sketch of whose head is given, (17). Very small, deficient intellect, &c., in the skull of a Spanish pirate, (18,) who committed suicide in prison at Cincinnati—a most desperate outlaw.

Physiognomical expression—this sentiment when very large imparts a mild, soft, and pleasing expression to the whole face, particularly the eyes and mouth.

Uses—desire to promote the happiness and well-being of others; universal charity and love.

Abuses—Prodigality; extravagantly fond of assisting others; loving, kind and indulgent to an excess.

Location—on the top of the head, immediately above the forehead.

> "The quality of mercy is not strain'd,
> It droppeth, as the gentle rain from heaven
> Upon the place beneath: It is twice bless'd;
> It blesseth him that gives, and him that takes.
> 'T is mightiest in the mightiest; it becomes
> The throned monarch better than his crown.
> His sceptre shows the force of temporal power
> The attribute to awe and majesty,
> Wherein doth sit the dread and fear of kings;
> But mercy is above this sceptre's sway,

40

No. 17. Miss ———— a Beauty of New York; a lady of exceeding kindness and generosity.

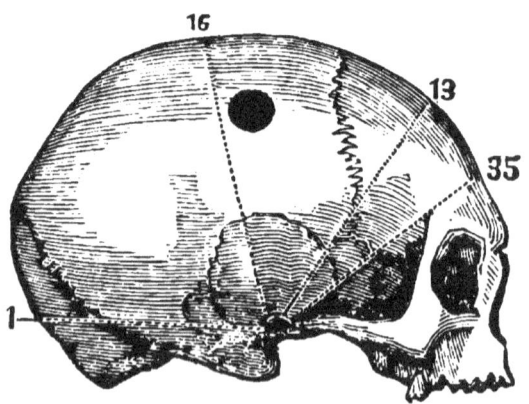

No. 18. Skull of a Spanish Pirate who committed suicide. Very deficient in organs 13, 16, 35.

No. 19. Veneration Illustrated.

It is enthroned in the hearts of kings;
It is an attribute to God himself;
And earthly power doth then show likest God's
When mercy seasons justice."

14.—VENERATION.

Very Large—Profound feeling of awe; reverence for the Deity
Large—Due consideration; respectful and kind to superiors.
Full—Little solicitude about religions matters; want of adoration.
Moderate—Inactive feelings of devotion; want of humility.
Small—Rude and overbearing; want of proper respect to superiors.

Veneration very large in portrait 19, of a very young miss, of New York, sister to the preceding lady, remarkable for her beauty, gentleness and goodness—contrasted with inferior skull, No. 20, very deficient in all the above.

Physiognomical expression—imparts a beautiful soft light to the eyes, as they are directed heavenward, the lips half severed, breathing the language of holiest love and calm devotion.

Uses—produces the sentiment of adoration and humility, and brings us in communion with God.

Abuses—immoderate fondness for ancient errors and superstitions, time-honored abuses, &c.

Location—behind Benevolence and before Firmness, on the longitudinal suture.

Sweet innocent! Her eyes upturned to heaven,
Do seem to seek their native home, the skies!
How soft, how lustrous, and how beautiful!
Say, does she see a heavenly seraph there,
Like her, all beauty, smiles, and loveliness?
Her roseate lips apart like opening flowers,
Inhale perfumes from heaven alone derived:
Blest creature! she is indeed an angel-child.
Her open forehead and her sunlit eyes
With radiant lustre shone, reflecting happiness
And innocence within. She was a fair, a gentle child,
So full of mirth and pleasantry, yet seldom wild;
Her hair in richest auburn tresses shone,
Dishevell'd o'er her alabaster shoulders hung,
In silken dalliance, with the gentle zephyrs playing,
Formed beauteous waving lines, like autumn's ripening fields.
Sweet fairy! I have listened to her laughing prattle;
How joyous, free, how gay and happy have I seen her!
Like some gay, carolling bird of jocund spring
Discoursing nature's heavenly music wild.

Her tiny feet so small did scarcely touch the ground,
It made her seem so little of this earth;
Imagination paints her now, a being of superior birth,
Almost too beautiful for this dull earth.
    Delightful, charming Ella! has she not forgotten me.?
Can I forget her?—ah! no, never!
Link'd, interwoven in fond memory's chain
With her fair, beauteous sister, I must remember, yes, forever

15.—FIRMNESS.

Very Large—Invincible perseverance; stubbornness and wilfulness.
Large—Great decision and resolution of character and purpose.
Full—Steadiness, stability; little disposition to change of purpose.
Moderate—Inconstant, wavering, and changeable; easily gives up.
Small—Variable, shifting, and easily abandons former views.

Firmness, large in Dr. Gall, (No. 21,) combined with the highest intellect, which enabled him to discover, appreciate, and establish Phrenology amidst a host of adversaries.

Physiognomical expression—compression of the lips; dignified, erect carriage; keeps on a straight course, regardless of the sneers of others. Very deficient in the skull of Peruvian, Idiotic Child, (No. 22.)

Uses—moral courage, steadiness, and determination.

Abuses—obstinate, unyielding stubbornness.

Location—the highest part of the head, and posterior to a line drawn perpendicular from ear to ear.

    "Though perils did
Abound as thick as thought could make them, and
Appear in forms more horrid, yet my duty,
As doth a rock against the chiding flood,
Should the approach of this wild river break
And stand unshaken yours."

16.—CONSCIENTIOUSNESS.

Very Large—Strictly honest and just; sensitive to the rights of others.
Large—The highest regard for truth, justice, probity and honor.
Full—Good share of integrity, but not over-sensitive in trading.
Moderate—No compunction for sin, and makes a close bargain.
Small—Will steal and lie; no scruples of conscience or honor.

Conscientiousness very large in the Beautiful Bostonian, (see title-page,) also in Charles Dickens, or "Boz," (No. 24,) the highly popular and talented friend of the suffering poor. Very small, with deficient Causality, in the skull of the New Zealand Chief, a cannibal. (See cut 25.)

44

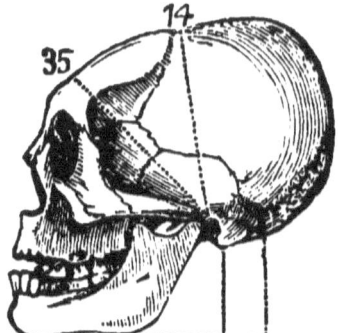

No. 20. Skull of a New Hollander. Small 14, Veneration, and 35 Intellect.

No. 21. Portrait of the great Dr. F. J. Gall.

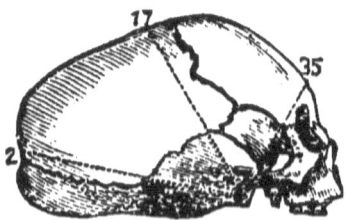

No. 22. Peruvian Idiotic Child. Very small 35 and 17

No. 24. Charles Dickens, or Boz, the popular author The Shakespeare of Modern Literature.

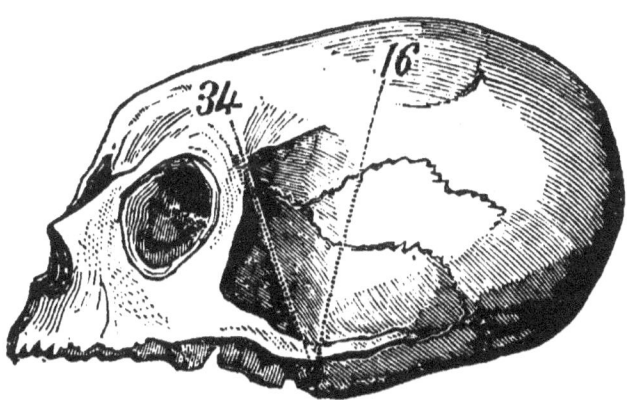

No. 25. A Cannibal Malay Chief. Very small 34 and 16; very large posterior. "Apply the same test of the lines to the superior heads."

Physiognomical language—amiability, openness of countenance, candid and sincere feeling, appearance of honesty.

Uses—to promote justice and love of truth.

Abuses—morbid sensibility at deriliction of duty in unimportant trifles; overwhelming feeling of self-abasement.

Location—adjoining and below Firmness, near the top of the head.

A ROGUE'S OPINION OF CONSCIENCE.

"I'll not meddle with it; it is a dangerous thing; it makes a man a coward; a man cannot steal but it accuseth him; a man cannot swear but it checks him; a man cannot lie but it detects him. 'T is a blushing, shame-faced spirit, that mutinies in a man's own bosom; it fills one full of obstacles. It made me once restore a purse of gold that by chance I found; it beggars any man that keeps it. It is turned out of all towns and cities for a dangerous thing; and every man that means to live well, endeavors to trust himself and live without it."

17.—HOPE.

Very Large—Extremely elevated spirits; prone to castle-building
Large—Vivacity and cheerfulness; anticipates great happiness.
Full—Apt to view the bright side of a picture; buoyant anticipation
Moderate—Reasonable desires, and not much ecstasy of feeling.
Small—Melancholy; depression; gloom; general despondency.

Hope very large in the lady on page 48, (No. 31.)

Physiognomical expression—this feeling imparts a vivacious, cheerful, and pleasing expression to the countenance, and adds buoyancy and elasticity to the person.

Uses—induces confidence in the future; support against disappointments and ill fortune.

Abuses—exaggerated ideas of happiness; chimerical, romantic, and absurd expectations.

Location—a little lower and nearly on each side of Veneration.

"But thou, O Hope! with eyes so fair,
What was thy delighted measure?
Still it whisper'd promis'd pleasure,
And bade the lovely scenes at distance hail.
Still would her touch the strain prolong,
And from the rocks, the woods, the vale,
She called on echo still through all her song;
And where her sweetest theme she chose,
A soft responsive voice was heard at every close,
And Hope, enchanted, smiled and waved her golden hair."

18.—MARVELLOUSNESS.

Very Large—Belief in the supernatural, witchcraft and demonology
Large—Credulous, fanciful, superstitious, and active imagination.
Full—Fond of the wonderful and astonishing; delights in romances
Moderate—Small degree of faith, want of credulity, not superstitious
Small—Very sceptical; distrusts even the best friends; incredulous

Marvellousness, very large in credulous persons.

Physiognomical expression—eyes and mouth wide open, ready to catch every new idea, which with such persons is swallowed with avidity.

Uses—a disposition to receive truth; fond of novelties and new ideas; the organ of faith.

Abuses—superstitious dread of ghosts, supernatural agency; credulous; easily imposed on.

Location—on each side and rather before Veneration, close to Hope, and behind Imitation.

> "The lunatic, the lover, and the poet,
> Are of imagination all compact:
> One sees more devils than vast hell can hold—
> That is the madman. The lover, all as frantic.
> Sees Helen's beauty in a brow of Egypt."

19.—IDEALITY.

Very Large—Poetic and ideal; gorgeous fancy; admires sublimity.
Large—Brilliant and excursive imagination and poetic fervor.
Full—Fancy, taste, and elegance; keen perception of the beautiful
Moderate—Absence of poetic taste and talent; not any refinement.
Small—Vulgarity and coarseness; want of elegance and sentiment.

Ideality very large, with the highest intellectual powers, in the immortal Shakespeare, (32,) contrasted with the skull of a remorseless cannibal, (33,) or Malay of St. Vincents, some of whom it is reported eat their victims alive. [See Appendix.]

Physiognomical expression—The poet, of all other beings, is perhaps the most sensitive, and is easily recognized by a romantic, visionary, and imaginative appearance and deportment

Uses—produces taste, imagination, and poetry, the beau ideal, excellence, the beautiful and grand.

Abuses—dreaming, abstracted and visionary schemes.

Location—this organ is above Constructiveness, and below Imitation, giving a fullness and squareness to the upper sides of the head, in the frontal and anterior region.

No. 31. Natural Language of Hope

No. 32. Portrait of Shakspeare.

No. 33. A Cannibal New Zealand Chief. Deficient in 19, and all the Intellectual Organs.

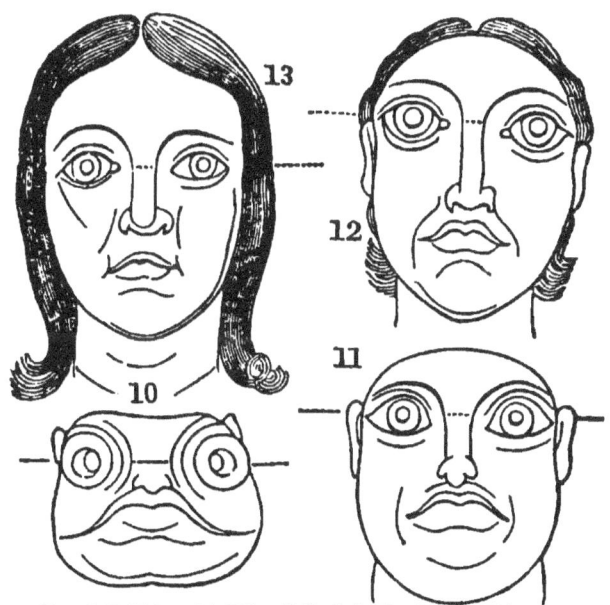

PROGRESSIVE SCALE OF BEING.

Nos. 34, 35, 36, 37; commencing with Frog, &c. The line through the eyes exhibits the proportion of brain to face. Apply the same test to the human head.

The poet's eye, in a fire frenzy rolling,
Doth glance from heaven to earth, from earth to heaven,
And gives to airy nothing
A local habitation and a name.

SHAKESPEARE'S INVOCATION TO THE MUSE.

"O, for a muse of fire, that would ascend
The brightest heaven of invention!
A kingdom for a stage, princes to act,
And monarchs to behold the swelling scene!
Then should the warlike Harry, like himself,
Assume the port of Mars, and at his heels,
Leash'd in like hounds, should famine sword, and fire
Crouch for employment."

? SUBLIMITY—(not usually marked.)

20.—MIRTHFULNESS.

Very Large—Brilliant at repartee; witty; fond of the ludicrous.
Large—Gaiety, wit, and laughter; jovial, pleasant, and humorous
Full—Agreeable and facetious, without much original witticism.
Moderate—Serious and sober; seldom excited to merriment or wit.
Small—Dull and tedious; no perception of the witty or ludicrous.

The figures annexed (34, 35, 36, 37) show the entire absence of this sentiment and of Ideality in the lower order of being; it is also very deficient in low, inferior human heads.

Physiognomical expression—laughing, merriment, hilarity, and cheerfulness of temper.

Uses—promotes sociability. Wit and ridicule may also be powerful allies in the cause of virtue.

Abuses—keen, sarcastic, cutting, envious remarks; ridicule of the great and good, &c.

Location—on outer parts of the top of forehead, giving it a squareness, as in Laurence Sterne.

"And jocund laughter holding both his sides.
                Now, by two-headed Janus,
Nature hath framed strange bedfellows in her time;
Some that will evermore peep through their eyes,
And laugh, like parrots, at a bag-piper;
And others of such vinegar aspects,
That they'll not shew their teeth, in way of smile,
Though Nestor swear the jest be laughable."

21.—IMITATION.

Very Large—Great talent for mimicry, caricaturing, or ridiculing.
Large—Cleverness to imitate either the mechanical or the fine arts
Full—Respectable talent only for imitating things; not a mimic.
Moderate—Inability to copy or act out; dislike to imitate any one.
Small—Original and eccentric in manners; failure to copy.

Imitation, very large usually in mimics, drolls, &c., and must also be very large in the monkey tribes. [See cut 39.]
Physiognomical expression—grimace, monkeyism, dandyism, &c.
Uses—to enable us to assimilate with others.
Abuses—to ridicule the great and good.
Location—on each side of Benevolence.

> "Monkey, little merry fellow,
> Thou art nature's Punchinello:
> Full of fun as Puck could be;
> Harlequin might learn of thee!
>
> Look now at his odd grimaces!
> Saw you e'er such comic faces?
> Now like learned judge sedate;
> Now with nonsense in his pate.
>
> There the little ancient man
> Nurses as well as nurse he can!
> Now good-bye, you merry fellow,
> Nature's primest Punchinello!"

ORDER 2—GENUS 1—Intellectual Facultie
Which perceive Existence and Physical Qualities.

22.—INDIVIDUALITY.

Very Large—Great talents for observation and critical judgment.
Large—Acute perception of everything seen passing around us.
Full—Desire to see and become acquainted; facility of acquiring.
Moderate—Absence of the noticing, observing, and retentive powers
Small—Want of observation; very deficient in noticing minutiæ.

Individuality very large, combined with very large intellect and sentiment, gives desire for and appreciation of beauty, as shown in the great sculptor Canova, (cut 39.) This organ gives acuteness of perception and ready talents, an aptitude to seize and combine the useful and the beautiful in nature or art. This faculty, when very large, imparts a strength of judgment, and general talents of a highly useful character, and is large in most distinguished men.
Location—at the bottom of forehead, between the eyebrows.

No. 38. Canova, the great Sculptor.

OURAN-OUTANG.

No. 39. Commonly called the Wild Man of the Woods.

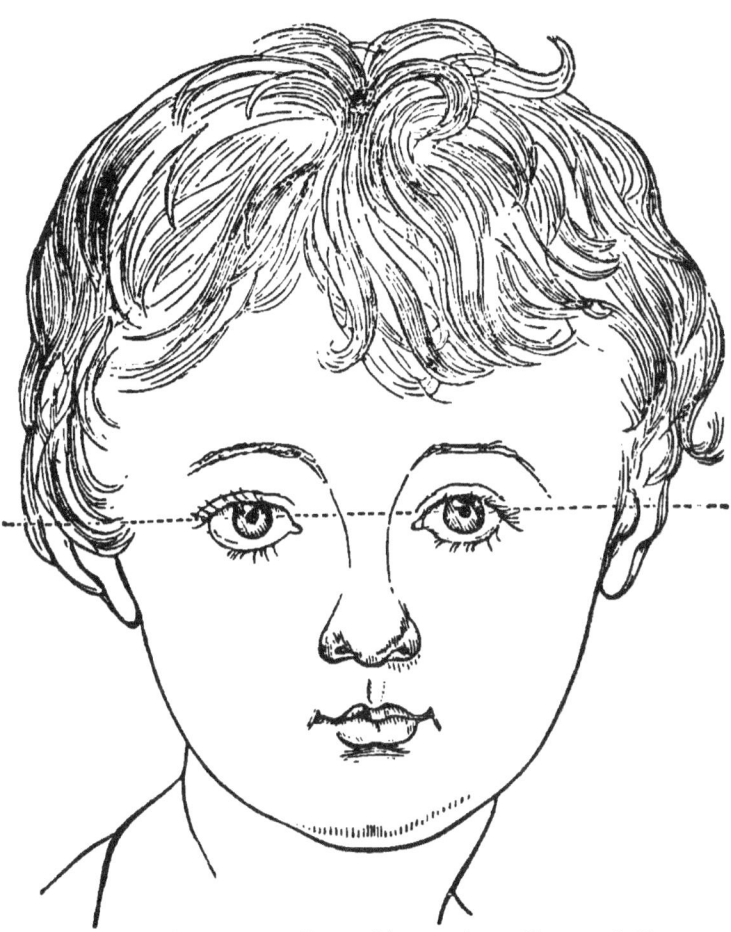

No. 39½. Precocious Boy, with very large Form and Constructiveness. The line drawn through the eyes is to show the proportion of brain to the face, which line ought in all well balanced human heads, when so drawn, to show as much brain above it as face below it. See the Idiot.

23.—FORM.

Very Large—Accurate perception of outline; talent for drawing.
Large—Power of delineating shapes; distinct memory of persons.
Full—Tolerable correctness of memory; moderate skill in drawing
Moderate—A weak memory, and indistinct perceptions of form.
Small—Absence of noticing external appearances, incorrectness.

Form very large in precocious boy of [See cut 39½.] This organ is invariably large in distinguished artists, sculptors, &c., and is highly necessary to mechanics, milliners, &c. In children, also, this organ, when large, greatly facilitates education, in acquiring the rudiments, forming and learning letters, &c.

Location—between the eyes; its size is estimated by the distance or amount of brain between them.

24.—SIZE.

Very Large—An excellent judgment of proportion, length, &c.
Large—Correct admeasurement of magnitude, space, distance, &c.
Full—Fair perception of size, length, and proportion of objects.
Moderate—Fails in guessing sizes; indifferent talents for surveying
Small—An extreme deficiency in estimating correct proportions.

Size, ability to judge of dimensions, space, &c., in proportion to the development of the organ. This faculty greatly assists the geometrician, mechanician, and engineer.

Location—on the lower side of the internal angle of the eyebrow, adjoining Individuality.

25.—WEIGHT.

Very Large—Intuitive knowledge of gravitation, momentum, &c.
Large—Very correct judgment of forces of bodies, preponderance.
Full—Facility in estimating or trying the weight of anything.
Moderate—Deficiency in balancing equilibrium, and in dancing.
Small—Absence of the talent of the discriminating of weights.

Weight, appreciation of momentum, resistance; also aids equilibrium, balancing, &c. It is necessary in engineering, hydraulics, mechanics, &c., to marksmen, musicians, tumblers, dancers, &c.

Location—outwards from size, and under the eyebrow.

26.—COLOUR.

Very Large—Great memory, judgment and fondness of colours.
Large—Talents for analyzing and harmonizing shades of colour.
Full—Accuracy in judging the effects and combinations of colours
Moderate—Want of talent, and deficient in skill for painting.
Small—Total absence of the faculty of distinguishing colours.

Colour, when large, implies correctness and facility of judgment in colours, painting, &c.; of harmonizing delicacy of tints in shading with colours, &c.

It is a well-attested fact, that great numbers, with the keenest sight, are unable to distinguish the great difference in even the primitive colours. This offers one of the most conclusive evidences of the existence of primitive faculties and natural endowments. [See the cut No. 40 of the organ in Peter Paul Reubens, the great painter.]

27.—LOCALITY.

Very Large—An extraordinary faculty of recollecting places.
Large—A good memory of localities, and fondness for travelling.
Full—A ready perception of localities, and does not easily get lost.
Moderate—Indistinct memory of positions, and soon becomes lost.
Small—Very deficient in memory of places, localities distances, &c.

Locality, if very large, great fondness for travelling, and remembrance of places seen. This organ is very large in celebrated travellers—Capt. Cook, Humboldt, &c. [See cut 41 of the organ in the former traveller.]

Location—on each side and a little above external of Individuality.

28.—CALCULATION.

Very Large—Intuitive perception of numbers; a skilful reckoner.
Large—Command of figures, and great facilty in computing sums
Full—Talent for figures, but not fond of exercising the faculty.
Moderate—Dislike to arithmetic and accounts, deficient in figures.
Small—Very slow and inaccurate in computing, reckoning, &c.

Calculation, power of computation, very large in Buxton, Bidder, &c., who, of ordinary minds in other respects, have astonished the world by their incredible powers of calculation; whilst some savage tribes of men are unable to count one hundred. [See cut 42, of J. Buxton.]

Location—outside external angle of the eye, next to order.

40. Organ of Color, very large.

41. Locality, very large

42. Calculation, very large.

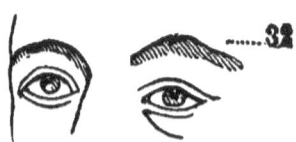
43. Tune, very large.

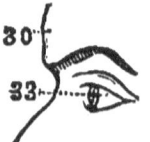
44. 30, Even., 33, Lang., very large.

45. Langurge, very large.

Josephine, the beauteous and beloved wife of Napoleon, and the origin of his good fortune, but who was wickedly sacrificed to his inordinate ambition. From the day of his divorce, his good fortunes deserted him, and at last he died a wretched exile on a barren rock. So, ambitious, soulless men, beware how you treat your wives.

5٩

"He left his country for his country's good."

No. 45½. Napoleon, the Emperor, who ignobly fought to aggrandize himself. By the power he acquired he could have liberated Europe by selfish ambition, he perished miserably.

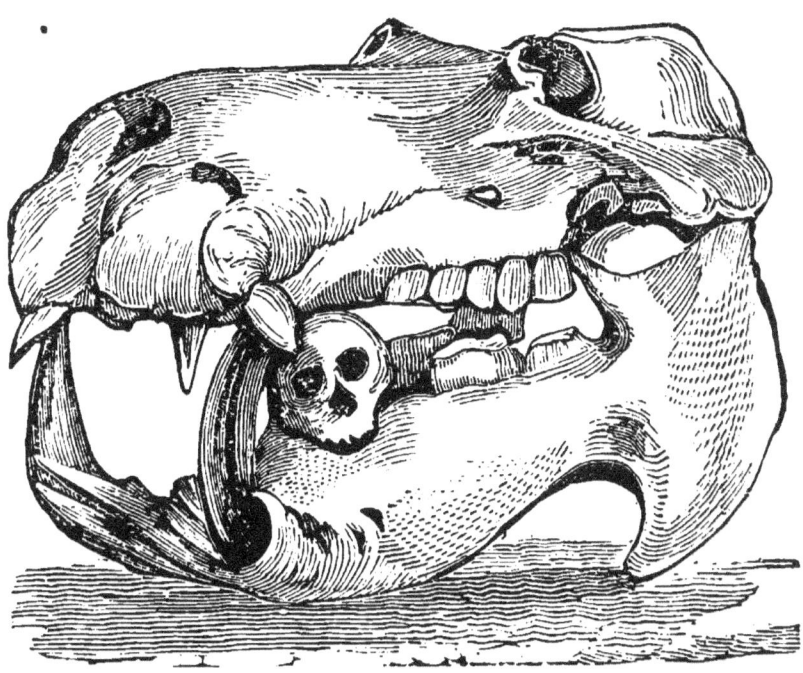

The bones of the head of the Behemoth, or Hippopotamus, described in the book of Job. The skull proper, or cavity which contained the brain of this monster, is not so large as that of the human skull, placed within its enormous jaws to show its size by contrast.

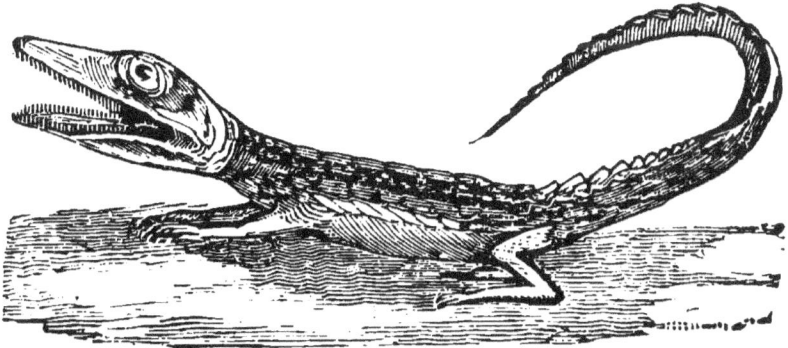

young Alligator from South America, with enormous Destructiveness, no lence, and without a particle of the Moral Sentiments—(all mouth.)

29.—ORDER.

Very Large—Extremely precise, particular, fidgety, and nice.
Large—Very methodical, systematic, and exact; great formality.
Full—Habits of order, but not very particular or attentive to detail
Moderate—Little precision or exactness, and a want of system.
Small—Confusion and disorder; general want of management.

Order implies the desire to systematize, arrange, and classify objects. Classification and generalization are absolutely necessary to the philosopher and successful man of business
Location—on the external angle of the eye.

30.—EVENTUALITY.

Very Large—Great powers of memory, and facility of acquiring.
Large—Easily acquires and retains knowledge; good memory.
Full—Power of remembering leading events, but not the minutiæ
Moderate—Inability of retaining much; a weakness of memory.
Small—Extremely forgetful of incidents or facts in the detail.

This faculty, when very large, (see cut 43,) enables us to treasure up whatever may occur—what we hear, see, or read. It may be said to be the power of recalling ideas to the mind, and is an essential element to success in almost every pursuit. Many can repeat, *ad infinitum, verbatim et literatim,* that which they have heard or read but once; whilst others have been known to forget the names of their most intimate friends, and even of their family, many amusing anecdotes of which are on record. We remember a case of a very reverend gentleman, who forgot his own wedding-day, and lost a wife. It was this defect in the great Dr. Gall first drew his attention to facts in nature, which resulted in the discovery of the science of Phrenology.

31.—TIME.

Very Large—Clear and correct ideas of time, memory of dates. &c.
Large—Accurate remembrance of chronological events and history.
Full—Indistinct notions of the lapse of time; a poor recollection.
Moderate—Incorrect as to dates; unable to keep or guess time.
Small—Extremely deficient and forgetful of dates or periods.

Time, perception and remembrance of chronological events, dates, &c. This faculty exists in very different degrees in various persons. Some have great facility in measuring, guessing, beating time, &c., whilst others are perfectly incompetent to do either. It is large in eminent musicians.

Location—adjoining and outside Locality.

32.—TUNE.

Very Large—Great intuitive fondness and ready talent for music.
Large—Musical taste and judgment, and a great lover of harmony.
Full—Extremely fond of music; good conception of melody.
Moderate—Mediocrity of taste in music; deficient in talent or skill.
Small—No appreciation of the science; inability to learn music.

Tune, (see cut 44, Handel, very large,) perception of melody, harmony, or discord. We have the most conclusive evidence of the existence of this organ, as it will be found that some who hear equally well with others, are quite unable to comprehend the simplest airs, or enjoy any of the pleasures arising from the *concord of sweet sounds;* whilst some again are so highly endowed with this faculty, they devote their whole existence to the pleasures of music. Some of the higher order of animals, as the horse, &c., have some endowment of this faculty, from the excitation and pleasure evidently afforded them by hearing music.

There is great difficulty in designating the exact strength or development of this organ, from the temporal muscle and ridge covering it. We therefore usually omit marking it, unless very large or very small.

Location—immediately above Number; indicated by width of lower temples.

THE MUSIC OF NATURE.

How sweet the moonlight sleeps upon this bank!
Here will we sit and let the sounds of music
Creep in our ears; soft stillness and the night
Become the touches of sweet harmony.
Sit, lov'd one; look how the floor of heaven
Is thick inlaid with patines of bright gold.
There's not the smallest orb, which thou behold'st,
But in his motion like an angel sings,
Still quiring to the young-eyed cherubim;
Such harmony is in immortal souls;
But, whilst this muddy vesture of decay
Doth grossly close it in, we cannot hear it.

"Whose battle-fields were holy ground."

No 46. The illustrious Washington, who only fought for his c untry's good and the liberties of all mankind. He is immorta .

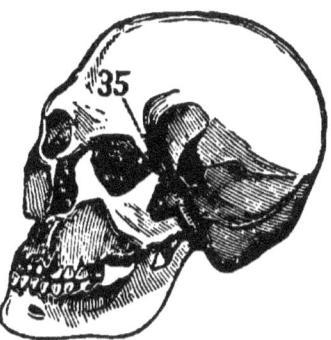

Nº. 164. The Skull of a Carib. Very deficient in 35, Causality, and large in the posterior or animal portion of the brain.

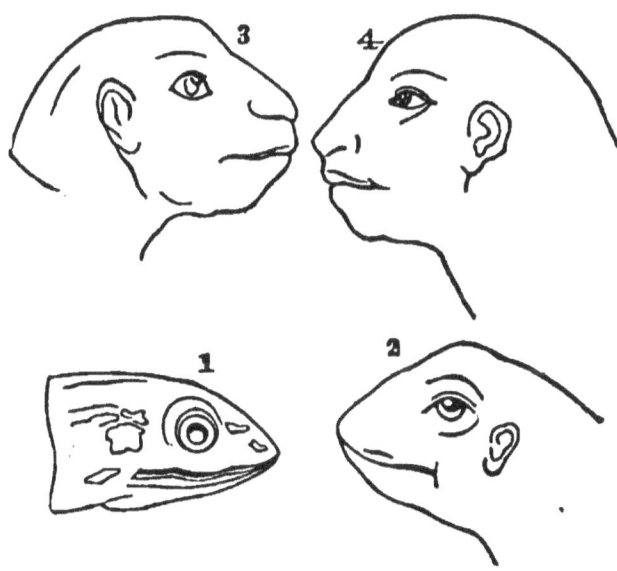

ANIMALITY.

Nos. 47, 48, 49, 50; exhibiting the head of Frog or Reptile, and gradual approach to the form of the head of Orang, as we ascend in the scale of organization. See page 106.

33.—LANGUAGE.

Very Large—Eloquent and ready in speaking; great flow of words
Large—Freedom of expression in conversing; free style of writing
Full—Not very communicative or loquacious on ordinary topics.
Moderate—Difficulty in conveying the correct meaning; bad style.
Small—Hesitating, embarrassed, deficient and awkward in speech.

Language, when very large, gives facility in communicating ideas by oral arbitrary sounds, or written signs; imparts facility in composition, and powers of rhetoric and oratory fluency in conversation and debate.

Uses—to communicate knowledge and promote sociability.

Abuses—garrulity, gossipping, and scandalizing.

Location—on the super-orbitary plate, and when very large depresses the eye outwards and downwards. [See example in Horne Tooke, figure 45.]

> "List his discourse of war, and you shall hear
> A fearful battle render'd you in music;
> Turn him to any cause of policy,
> The Gordion knot of it he will unloose
> Familiar as his garter; that when he speaks,
> The air, a charter'd libertine, is still,
> And the mute wonder lurketh in men's ears,
> To steal his sweet and honeyed sentences."

GENUS 2—Reflective Faculties.

34.—COMPARISON.

Very Large—Extraordinary talents and strong critical judgment
Large—Correct powers of analyzing, comparing and criticising
Full—Fair judgment, good practical talent, and a close observer.
Moderate—Tolerable skill, but not much clearness of perception.
Small—Superficial reasoning; no depth of thought or intelligence

Comparison, very large in Washington, (see cut No. 46,) combined with the purest moral sentiments, gave its possessor wisdom, sagacity, and judgment of the highest order, contrasted with skull of Carib, (46!.) Washington in his life exhibited those great and good qualities so harmoniously blended, that by common consent he is placed at the head of great and estimable men.

Uses—to discover and apply truths by analogy, force of reason; ability to investigate, to discover good from evil, truth from error, &c.

The whole range of the mental and physical world are subjected to this analyzing, scrutinizing, intellectual, thinking principle, by which man is elevated infinitely above all other beings,[contrasted with outlines of animality, 47, 48, 49, 50, in which may be comprised nearly the whole of animal or inferior heads, from frog to orang.] This reasoning faculty leads man to investigate the phenomena and wonders of the universe, and animated nature, "and so through nature up to nature's God."

Uses—to reason analogically; to compare closely; to reason *a priori*, (from a prior cause.)

Abuse—abstract intensity of thought, to the neglect of religious or temporal duties.

Location—it is situated in the highest part of the middle of the forehead.

SHAKESPEARE'S DESCRIPTION OF A GREAT MAN.

See what a grace was seated on this brow:—
Hyperion's curls, the front of Jove himself;
An eye like Mars, to threaten and command;
A station like the herald Mercury,
New-lighted on a heaven-kissing hill;
A combination, and a form, indeed,
Where every god did seem to set his seal,
To give the world assurance of a man.

35.—CAUSALITY.

Very Large—Great originality of thought, powers of invention.
Large—Energetic and active habits of mind extremely inquisitive
Full—Disposition for inquiry, but incapable of profound thought.
Moderate—Indifference to metaphysics; little or no inquisitiveness.
Small—Deficient in reasoning power, and weakness of intellect.

Causality implies the desire to ascertain, why, wherefore is this so? This organ is very obvious in the head annexed of the great American philosopher, Franklin, (cut 51,) to whose profound spirit of enquiry the whole world is deeply indebted for some of the most valuable and important truths in natural philosophy. It was this great philosophical spirit which sustained him, a wandering boy, with his roll of gingerbread under his arm, and which subsequently enabled him to persevere till he brought the lightning within his grasp, and rendered it subservient to the use of man.

This may be said to be "the divinity that stirs within us," or that God-like attribute which we call reason. Plato, Socra-

No. 51. Benjamin Franklin, the great American Philosopher, discoverer of Electricity, &c.

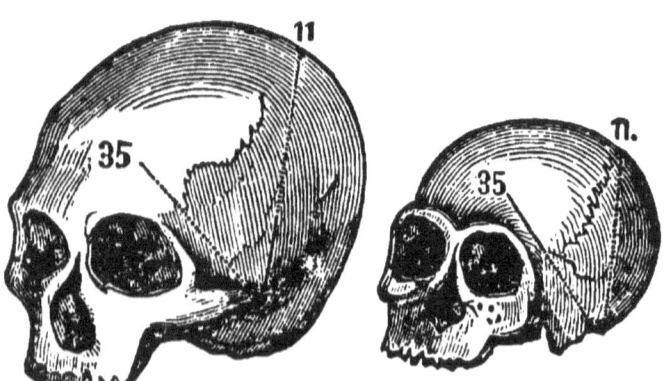

No. 51. The Skull of an Idiotic Girl, also of an Orang-Outang.

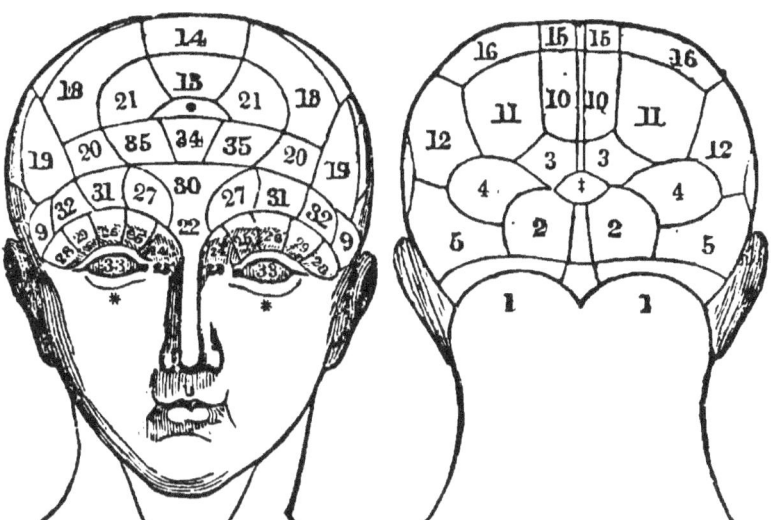

Nos. 56 57: showing the locations of the Organs on the back and fron of the Head.

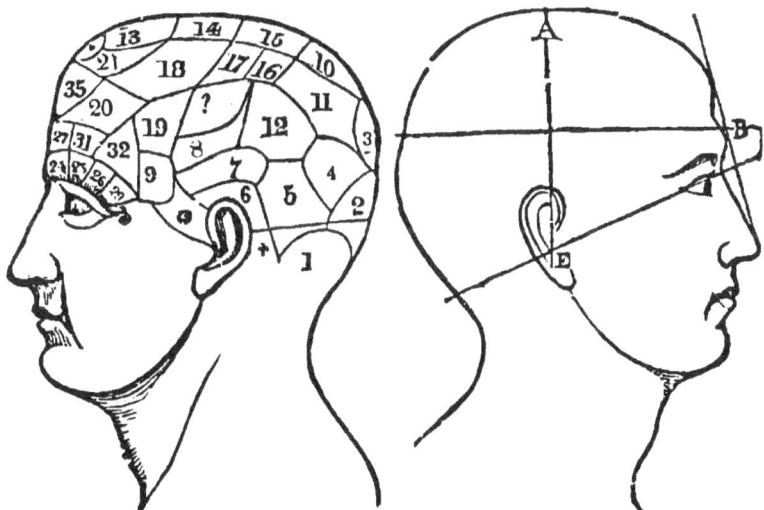

Nos. 58, 59. Figure 59 shows the general divisions of the Head explain on page 73.

tes, and other eminent philosophers, or friends to wisdom. were largely endowed with this faculty, which may be considered as one of the most distinguishing and noblest attributes of man.

Uses—to trace cause and effect; to pursue philosophical inquiry.

Abuses—metaphysical subtleties without a practical application.

Location—outside and each side of Comparison.

Physiognomical language—serious, thoughtful and contemplative.

INTELLECTUAL AND SENTIMENTAL BEAUTY DESCRIBED.

Two of far nobler shape, erect and tall,
God-like erect, with native honor clad.
For contemplation he and valor formed,
For softness she and sweet, attractive grace;
He for God only, she for God in him.
His fair, large front and eye sublime declar'd
Absolute rule; and hyacinthine locks
Round from his parted forelock manly hung
Clustering, but not beneath his shoulders broad;
She, as a veil down to the slender waist,
Her unadorned, golden tresses wore.

ADMEASUREMENTS OF THE HEAD BY TAPE.

			Coombs's Bust.
	Circumference of the base, close above the ears		22¼ inches.
	Circumference of head around the organs of Cautiousness, Causality and comparison,		22 inches.
From the opening of one ear to the other.	Over lower backhead, or the organ of Amativeness,		9¼ inches.
do. do.	Over perpendicular tophead, or the organ of Firmness,		14¼ inches.
do. do.	Over front tophead, or the organ of Benevolence,		13¼ inches.
do. do.	Over top of forehead, or the organ of Comparison,		12¾ inches.
do. do	Over lower forehead, or the organ of Individuality,		11¾ inches.
	Individuality to occipital spine,		3¾ inches.

THE TEMPERAMENTS.

Physiologists have laid down the following rules in forming a judgment of the temperament from the external appearance of the individual, which are described as follows:

THE LYMPHATIC is distinguished by a roundness of person, fair hair and skin, delicate texture of body, and softness of the muscles, inactivity of the brain and nervous system.

THE SANGUINE temperament is indicated by fulness of person and firmness of muscle, bright chesnut hair, ruddy countenance, and regular features; fond also of animated and active exertion.

THE BILIOUS temperament is recognized by full, dark hair, angular outlines of form, firmness of texture; also by strength and energy of person; the features acute and strongly defined.

THE NERVOUS temperament is distinguished by rapidity of motion for a short period; small muscles and thin, fine hair: easily becomes fatigued, and very susceptible.

When comparing different brains, the temperament should always be attended to; because two brains may be of the same size, but if one be of the lymphatic, and the other of the bilious temperament, there will be great difference in the powers of manifesting the faculties. In referring man's actions to his cerebral structure, we must admit the great importance of his organic constitution or structure, as this is one of the first and most important conditions to be observed in estimating his phrenological character. We can readily conceive how the organic constitution of the brain is affected and modified by the digestion, circulation, perspiration, and nutrition of the body, and how these different states of organization may produce different degrees of activity of the mental faculties generally.

The size of Brain, other conditions being equal, is the measure of power, either in the aggregate or as exhibited in detail by means of the written figures in this book, which, as before stated, are intended to express the relative size or force of each particular organ or faculty.

VERY LARGE.—A person having a Head or Brain marked VERY LARGE, with activity also VERY LARGE, with a favorable temperament, good education and opportunities, will exhibit the very highest order of talents and vigor of intellect, will be a natural genius and possess an aptitude for almost any pursuit or profession, and calculated to make a figure in the world. He will, by the mere force of his natural genius, be enabled to overcome difficulties which would be insurmountable to persons of smaller heads and ordinary talents; such an one will shine like a bright and particular star in the firmament of intellectual greatness, upon which future ages may gaze with astonishment and delight. His mental capacity will grasp the highest, the noblest, and the most sublime conceptions of happiness and virtue; his feelings will be of the most exquisite sensibility, either for pleasure or pain. With activity LARGE, he will be endowed with a very high order of talents and very superior powers of mind and vigorous intellect, enabled to make a distinguished figure amongst even great men, and be a leading, prominent character in whatever sphere he may be engaged. With activity FULL, on great occasions, or when thoroughly aroused, he would be truly great; but upon ordinary occasions he would not manifest those traits of character which would distinguish him from the generality of mankind.

LARGE.—One having a large-sized Brain, with activity LARGE or VERY LARGE, will possess great power of mind, and exercise a commanding influence over those with whom he may come in contact. He will possess great ardor and impetuosity, and in some points be irresistible, particularly should his propelling powers and selfish passions be strong. With the practical intellect LARGE, he would excel in business, and possess a ready intuitive knowledge of everything, and be highly successful in his profession or business. With activity FULL, he would be endowed with a great share of mental capacity, but require stimulus to exertion and thoroughly arouse him to those vigorous efforts of mind, of which he is under some circumstances capable; when not excited, he would pass for an ordinary person. With deficiency of the perceptive and reflective faculties, would not be very shrewd or apt, but rather inferior in judgment and capacity; but with large propelling or animal feelings, might exhibit a great degree of misapplied energy, and a great wish to excel, but not the capacity. With activity MODERATE, when powerfully excited, may evince considerable energy of intellect and capacity for performance, yet be too indolent and sluggish to do much; wanting also in clearness of ideas and intensity of feeling, and unless driven to exertion, will not be likely to accomplish much. With activity SMALL, or VERY SMALL, an extreme aversion to action, indolent and incapable of great exertion, either mentally or physically.

FULL.—With activity GREAT, or VERY GREAT, and the practical intellect and propelling powers LARGE, or VERY LARGE, although not possessing a high order of talent, will be generally clever, have considerable talent, and that so energetic, that it will pass current for more than what it really is worth, from its great incentive to action, yet is inadequate for great undertakings, and incapable of making a distinguished figure in the world, nor be really great. With activity FULL, will be tolerably clever, but manifest only an ordinary share of intellect. With the intellectual and perceptive faculties LARGE, he would be enabled to conduct business of an ordinary character, and even to obtain some celebrity and pass for a talented man. With activity MODERATE, will be indisposed to action, and take the world easy. SMALL, or VERY SMALL, will be extremely deficient in everything that constitutes a great character.

MODERATE.—One with a Head of only moderate size, combined with GREAT or VERY GREAT activity and large perceptive and propelling powers, will appear to possess much more talent than he really does, and with others to pioneer for him, he may be enabled to follow their plans with advantage. He will be more remarkable for activity than strength of mind, and more showy than useful in his acquirements. With activity only FULL, will have but a very moderate amount of intelligence, and no desire to acquire a great reputation; very sluggish and inactive. With activity MODERATE, or SMALL, extremely dull of apprehension and excessively indolent.

SMALL, OR VERY SMALL.—One with a SMALL or VERY SMALL Head, will be conscious of little else than merely an animal existence, and can never accomplish those high and splendid achievements which have immortalized the names of Homer, a Milton, or Shakespeare, or of the super-eminent goodness or greatness of an Alfred, or a Washington, who, generously resigning every consideration of self for the advancement of their country's happiness and welfare, have left their names an enduring monument to all ages.

THE DEGREE OF ACTIVITY.—Whilst size gives power, momentum and endurance of the mental faculties, &c., activity imparts intensity, quickness, willingness, and even a restless desire to act; implying energy and efficiency of character in the same ratio as marked in the margin.

The Brain is divided into two halves or hemispheres, each hemisphere being composed of a number of folds or convolutions, each of which has been discovered and unfolded by Spurzheim's new and admirable method of dissecting the Brain. The functions of each have also been fully established by the unwearied and indefatigable personal observation of the immortal Drs. Gall and Spurzheim, first founders of the science, and since corroborated by innumerable practical observations of its numerous professors, both in this country and Europe. The two hemispheres of the Brain are brought into communication and simultaneous action by means of fibres running transversely from one to the other. This important fact was first established by the extremely delicate method of unravelling the Brain, as practised by Spurzheim. The Skull, or bony outside covering to the Brain, and its three distinct coverings by which it is enveloped, generally presents a perfect parallelism to the Brain, as it is moulded on the Brain, and may be regarded as a kind of shell-work, as it presents no more obstruction to the growth and development of the Brain, than does the shell to the growth of such animals as are protected by them; the bony structure being of a secondary formation to that of the Brain or softer parts of the human system.

ON PHYSIOGNOMY.
DEDICATED TO THE LADIES.

In order to invest this subject with more interest, each of the passions, sentiments and intellect are described in verse, in order to exercise the higher order of faculties, also to render their action and influence on the character more conspicuous and the more easily retained in the memory. The author hopes he has also added to the attraction of this very interesting subject, by describing the various emotions of the mind, as transitorily depicted on the countenance, and which, undoubtedly, if long or habitually indulged in, permanently leaves its impress there; from which, no doubt, persons largely endowed with the perceptive faculties, are enabled to form a very correct estimate of the leading passions by a close observance of the features. Hence we conceive the science of Physiognomy has originated, and which is undoubtedly the key to the leading passions and feelings of individuals. We are, however, far from supposing it can be reduced to any systematic mode of investigating character. From its wonderful and beautiful mobility and ever-changing expression and variableness, it may be compared to the fleeting summer cloud—

> Now lighted up with heavenly azure brightness,
> Anon dark, driving clouds and tempests lower,
> And sheeted lightnings rend earth's fairest flowers,
> And ruin stalks abroad to desolate the land:
> So most wondrous beauty (the more 's the pity,)
> May be transformed, with vengeful ire,
> To frightful rage and horrible distortion—
> Dread foes to peace, to friendship, and to love.

There is one singular fact, important to the ladies, particularly those who are desirous of preserving unimpaired the beauty with which heaven has favored them : it may not be uninteresting to learn, that public speakers and many others have declared that by certain expressions of the features the corresponding emotions of the mind are produced with a vivid intensity. This probably proceeds from the nerves, muscles, &c. which connect certain parts of the brain with corresponding parts of the face; so that ladies who wish to preserve this precious gem, their beauty, must ever indulge in the kind and gentler emotions, and avoid all irritation, both of look and feeling, as they would the Scylla and Charybdis, or the wreck of beauty and loveliness.

THE HUMAN BRAIN

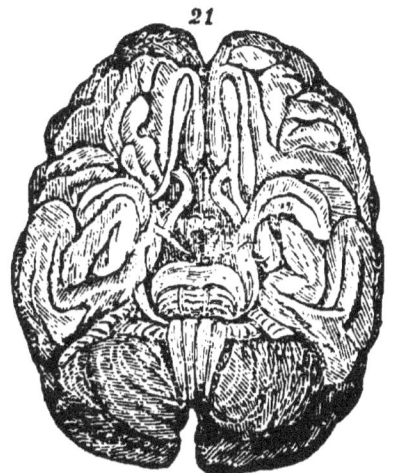

Lower View of Brain.

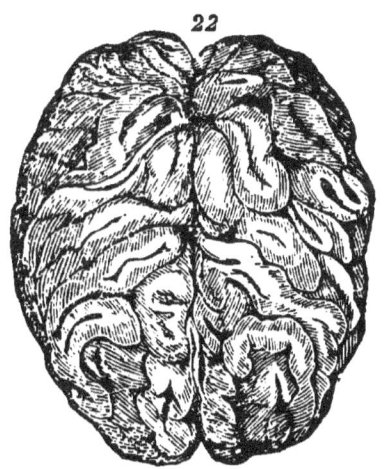

Upper View of Brain.

The preceding cuts are correct representations of the upper and lower view of various parts about the base of the brain of an adult intellectual person.

The brain is divided into two halves, or hemispheres, each of which are united by means of minute filaments or thread-like substances, embedded in cortical and medullary matter, radiating in various directions, crossing each other, communicating the two hemispheres of the brain, the Medulla Oblongata, the Medulla Spinallis, the various organs of sense, and all parts of the body.

The intercommunication of these remote nervous filaments with those in the brain is instantaneous. The nerves of volition and sensation act in the same manner.

The singular and extraordinary effects of electricity and galvanism on the human body, animate or inanimate, point to this subtle fluid as entering largely into the animal economy, the nerves appearing to act as the conductors of electricity, which moves with a velocity of four hundred thousand miles per second, or about the same velocity as light.

The brain proper is called the cerebrum, and the smaller brain the cerebelum. The former is much more voluminous and various in its form, structure, and functions, and wherein are located the organs of the various passions, intellect and sentiments; the smaller brain, or cerebelum, being solely the organ of physical love, and gives rise to the feeling of Amativeness, although it has recently been conjectured being also the seat of the organ of voluntary motion

The brain is protected by three distinct membranes or coverings, in which it is enveloped. The first is the *pia mater*, which closely adheres to the surface of the brain, dipping into the sulci, or cavities, and conveying innumerable blood-vessels to it. The second is named the *tunica arachnoida*, resembling a spider's web in fineness. It secretes a fluid to lubricate the *pia mater* and the *dura mater*, or third covering, which is attached to the brain, and also adheres to the inner surface of the skull.

The skull is curiously formed of eight separate bones—one frontal, two parietal, two temporal, one occipital, one sphenoidal, and one ethmoidal—each united by sutures or seams. The growth of these bones is each of them independent of the other, and commences growing from its own centre, and in old age firmly intersect each other by means of curious dove-tailed seams, which are much more serrated in the Caucassian skulls than the Carib, and still more simple in animals, &c.

The formation of the skull may be regarded as an excrescence, or shelly covering, of a secondary growth, (as seen in children.) It is designed for protection to the brain, and not to impede its growth, as some appear erroneously to imagine.

Previous to the anatomical researches of the founders of this science, the brain was supposed to be merely the root from whence originated the whole nervous apparatus which proceed from it to every part of the human structure, which led many distinguished anatomists and philosophers to conjecture that the brain was in some measure the sensorium, or seat of the intellectual powers; but nothing definite was known of its structure or functions until the discoveries of Drs. Gall and Spurzheim.

Our limits will not permit us here to show the singular correspondence of the mind, &c., with this its instrument, not only in the human family, but to the lowest order of being. The complication of structure, and large size or volume in man, orang, &c., is wonderfully contrasted with the smallness of volume and extreme simplicity of structure in the lower order of animals, reptiles, &c., being in the latter reduced to a mere point or particle of gelatinous matter, entirely destitute of convolutions, or appendages of nerves, &c., as we find in the higher order of beings.

To those desirous of pursuing this study in detail, we have much pleasure in recommending the splendid work of Spurzheim on the Anatomy of the Brain.

FOUR DIVISIONS OF THE HEAD.

The cut No. 59, with the line perpendicular from the ear, e to a, represents the anterior and posterior portion of brain, and the line through the eye and ear represents the base or foundation of brain by which we can estimate the proportion of brain to the whole head or face inclusive. The line from the nose to forehead gives the facial angle. The line from C to B divides the sincipital from the occipital region, the portion above being peculiar only to man, including the intellectual and sentimental or controlling organs—the portion below being the propelling or selfish propensities, which are common to man and the lower animals.

FOUNDATION OF PHRENOLOGY.

PHRENOLOGY is a derivation from the Greek, *phren* and *logos*, or the science of mind. This compound word is the adoption of Dr. Spurzheim, the distinguished associate of Dr. Gall. To the latter belongs the high honor of being the first to discover this new science of mind, or, in other words, the mode of ascertaining man's peculiar innate talents and original powers of mind by a reference to his cerebral organization, or form and volume of brain. Like most other exact sciences, it has been of slow progress, and has not been rashly adopted, but is the result of a most laborious, patient, and minute investigation of the human cerebral structure, in its endless and interesting varieties, both in states of perfect health and every stage of the various diseases to which we are subject. The human brain—that gordion knot, which has puzzled the sages of ancient and modern times, and which they could only *untie* by *cutting*—has now, for the first time, been completely unravelled, and its most wonderful beauty, complication of structure, and harmonious adaptation to its functions, been laid open by the labors of these distinguished physicians and philosophers.

Further; in order to satisfy the most incredulous, they have carried their indefatigable researches to those minute points in creation, or most simplified of animals and reptiles, and even to those minute animalculæ, the ephemeral existence of a day. The results of their investigations of human and comparative Phrenology have appeared in volumes of facts, sufficient to satisfy the most incredulous of the reality of this science, and of its high importance to the happiness and well-being of man. Their united labors were indeed conducted with singular ability, zeal, and enthusiasm, and which have comparatively left little to their successors but further to establish and confirm if possible their previous discoveries.

To the great Dr. Gall we owe the rude discovery of the science; and whilst as yet in its incipient stages of existence, he had both incorrectly named the doctrine itself and also several of its most important functions. This resulted, in a great measure, from the fact of his having discovered the several organs when in an excessive state of development, or as manifested in their abuses. As an instance, from the comparison of the heads of notorious, incorrigible thieves, although there might be many dissimilar forms of heads, yet in this one particular region of Acquisitiveness he found a very great enlargement. It was this fact which led him erroneously to suppose a particular faculty or organ of stealing. This was also precisely the case from an examination of the heads of murderers, by discovering the organ of Destructiveness enormously developed, and which he incorrectly named the organ of murder. On this account a stigma has attached to the science, which its opponents have made great use of. The science itself also was incorrectly named by Dr. Gall, being that of Cranioscopy, or Craniology.

We must here take this opportunity of observing, that none of the faculties, as discovered and located by Dr. Gall, have been ever materially changed. It is to the philosophic spirit of inquiry, and severe mode of inductive reasoning, united to the anatomical discoveries of the accomplished Spurzheim, that we owe its present correct classification and nomenclature.

APPLICATION AND UTILITY OF PHRENOLOGY

PHRENOLOGY is the only science of mind susceptible of actual demonstration, and is the only true philosophy of mental action It is a powerful support to morality and religion. It is the only rational and true basis of education. It is the foundation of domestic happiness. It is a most powerful auxiliary in the cause of morality, religion, virtue, and of education. It not only teaches us to know ourselves, but it goes a step farther, and teaches us to know others also. Its principles and practice are invaluable in jurisprudence, civil and criminal. By referring man's intellectual and sentimental innate character to his organization, it strongly appeals to our charitable sympathies to make allowances for the imperfections of others, and to lay no more upon individuals than they can bear. It also teaches us to desist from the mad attempts which have been made to reduce the talents and opinions of all to one standard.

Bell and many other distinguished physiologists and metaphysicians have admitted that the mind manifests a plurality of faculties. The discovery that it employs the brain as its agent, was left to immortalize Dr. Gall.

"Size of the brain, other conditions being equal, is a measure of power." This proposition is supported by analogy throughout nature, and by observation. The conditions to be observed are, 1st, Temperament; 2d, Age; 3d, Health; and 4th, Exercise. Phrenologists contend that if these conditions are equal in two individuals, the one possessed of the largest organs will manifest superior powers of mind. These conditions should be kept in view, also, when comparing different compartments of the brain; for one individual may have a smaller brain than another, and yet manifest the greatest amount of intellectual power. This will be seen to arise from the small brain being endowed with a larger intellectual lobe than the larger brain. But here the conditions vary, and the judgment of the two heads must vary accordingly.

The form and size of the brain, and each of its divisions, may be ascertained to a mathematical, actual certainty, from the external appearance of the head—while the health, &c., can be easily determined by other external indications.

IMPORTANT TO MOTHERS.

PHRENOLOGY teaches us that the brain is the organ of the mind, and also that the mind is not an innate or distinct single power, and capable of acting in every direction alike, and with equal force and intensity, but that it is multiplex in its operations, and is composed of a number of distinct organs or faculties, each organ being the name of a certain portion of brain, which organ is estimated by its size or volume (other conditions being equal.) Thus, the size of brain in the anterior and superior portion of the forehead of the individual is the measure of his intellectual capacity. The height of the head, or fulness in the coronal portion, indicates the strength of the moral and selfish sentiments; whilst the width of head gives us the estimate of his animal or selfish propensities. It will be seen by this classification, that Man possesses a Moral, an Intellectual, and an Animal Nature, each acting in concert and producing an harmonious whole, where each of these are properly balanced, as in the most perfect form of heads. Should the numbers range highest in the animal or selfish propensities, it must be inferred that the individual is mostly influenced by the lower or animal feelings, although more or less modified according to their relative proportions by the organs of intellect and sentiment. In the absence of the animal portion, the individual lacks energy and power. The general form of the head determines the character of the individual, and not particular bumps or depressions, as erroneously imagined.

In forming an estimate of character, it must be remembered there are several highly important conditions to be considered, which have great influence in the formation of character—the temperaments, education, circumstances, and habits of the individual. The purely lymphatic is the very least disposed for action, with great lack of energy in the brain and nervous system. With regard to their culture, or exercise of the organs, persons having had the advantages of a superior education, would exhibit a much greater degree of intelligence or sagacity than persons without those advantages, but whose heads should be precisely alike in every other particular of size, configuration, &c. Thus, again, men are much influenced by circumstances, and which most materially modify the character, and which will also unquestionably alter the configuration of brain or head by any course of conduct or study long and pertinaciously persisted in. It may be seen that children of superior or highly endowed persons will not only possess the exact form of head of one or both parents, but will also, from early habitudes, become almost counterparts of one or other of the parents; and under those favorable circumstances, such children are from the earliest infancy under a course of judicious mental cultivation, which, apart from all natural high endowments as to the form of crania, &c., must exert an influence on the future character, which would be almost incalculable for good. The very reverse of this is also precisely the case where low and dissolute persons have children under their control, and, it may be added, mismanagement. From such unfortunate associations the worst consequences may be apprehended, even with natural capacities of the highest order, and the finest form of crania. Children and persons with such associations would be more or less demoralized. The science is in this respect of such an immense value, it learns us that the formation of character of children, youth, &c., depends very materially on ourselves; for it must be observed, that the system of education, either for good or evil, commences almost in the cradle; and the influence of the mother on the formation of character is too much overlooked, and its importance too much lost sight of. The elegant but superficial acquirements which at present too much usurp the attention of young ladies, might in some measure be discarded, and the science of Phrenology, Physiology, &c., be substituted in their place, in order the more perfectly to qualify them to perform aright those high and important duties which, as mothers and the heads of families, usually devolve on them. It might be added, however, that American and European ladies, particularly in the higher circles, are many of them distinguished for their varied and highly useful scientific attainments, clearly proving, in many branches, nothing inferior to the boasted superiority of the male sex.

It has been aptly remarked, by Napoleon, that no great man ever had a weak mother; and undoubtedly his judgment was perfectly correct in assigning to the maternal side the greatest influence on the formation of character. On her alone hang the destinies of this republic, for children are almost the creatures of an intelligent mother's will. The phrenologist, in giving his estimate of character, will not, of course, be generally aware of all these foregoing important conditions, which so obviously and very materially affect the character. It must, therefore, be conceded that there are peculiarities of character, which legitimately do not come under the cognizance of this science, and for which due allowance will be made by the candid inquirer after truth. It is the paramount object of the science to point out particular excellencies and aptitudes of character, or the natural talents and disposition, also to point out defects, and to apply its proper corrective, by cultivating and exercising its antagonist faculties or opposing forces. Probably no condition is so necessary for the perfect possession of our faculties, and enjoying them in their highest degree of perfection, than a perfectly sound state of health, a compact and vigorous constitution, and energetic physical powers. Indeed, this is a consideration which has been so entirely overlooked, and the inference naturally drawn from it, that without a healthy physical organization, it is in vain to expect a vigorous, clear, and healthy exhibition of the mental powers. Many have yet to learn that man, however great his intellectual powers may be, can no more with impunity overtask the organs by which he exercises his intellectual powers, or sustain consecutive action of thinking for forty-eight hours without intermission, than can an individual exercise his physical or muscular energies for the same space of time, without great consequent loss of energy and exhaustion, alike injurious in both cases, and probably the delicate organ of the mind would be the greatest sufferer. Indeed, we have so many corroborations of this, in the premature deaths of so many bright and precocious geniuses, who have fallen early victims to over-exertion of the mental faculties in early infancy, we are in bounden duty compelled to caution all who have the charge of infants, or very young children, to be extremely careful not to over-task the tender and delicate organs of the mind. It would be extremely absurd to expect them to exhibit extraordinary mental acquirements, or very great physical energy, before their brain or physical structure has attained a perfect form and consistency. The first consideration ought to be the establishing a sound and vigorous constitution, as on this will greatly depend their future happiness and welfare, and which an affectionate mother only can duly estimate.

UNDENIABLE PROOFS OF PHRENOLOGY.

Phrenology has been established and rests its claims to support solely by an appeal to facts, by actual demonstration and by induction. This rigid mode of establishing the science invariably leads to the conviction, that,

1. We have no evidence of thought or mind without brain: we therefore affirm it to be the organ of the mind, or the instrument and *modus operandi* by which the intelligent principle carries on its operations.

2. Because, by anatomical researches, we invariably discover, in the endless chain of beings, the brains of men and animals to be precisely in accordance with the various peculiarities, dispositions, and talents they are known to possess. In men only of superior formed heads do we find large cerebral masses or volume of brain in the forehead or frontal, also in the coronal or superior portion, giving an innate feeling of justice, veneration, and charity, also the peculiar aptitude for poetry, painting, mechanism, and metaphysics, combining the highest order of intellectual pursuits.

3. Because in men we find an infinitely greater diversity of forms of head or brain than in any other created beings, of the same species, whatever. In man, also, we find an infinitely greater diversity of talents, sentiments, and feelings, singularly corroborative of the phrenological analysis of his nature.

4. Idiocy is incontrovertibly proved to result from a deficiency of the intellectual organs, or from disease of that particular portion of brain, the forehead. Partial insanity, or mental hallucination upon some subjects only, is ascertained, by *post mortem* examinations, to be the effect of either external or internal injuries of those portions of brain only which are affected. Dreaming also results from imperfect sleep; those portions of the brain in this case retaining a degree of consciousness, by which we afterwards recall some of these impressions. Perfect sleep being perfect unconsciousness, man, in this state, closely resembles vegetative existence.

5. Because the heads of infants and adults are both extremely dissimilar, and both strikingly illustrative of their characters. The very differently formed heads of the opposite sexes, but of the same variety or even family, are always in accordance with their various known characteristics.

6. The hereditary transmission of family peculiarities, talents, and dispositions, are in some cases strikingly singular, and can be accounted for on no other than the phrenological principle, or the correspondence and similarity of organization of the brain, form, and features—"the like producing its kind." Human action is clearly referable to organization. "modified by circumstances,'

temperament, and education. Human conduct is totally inexplicable upon any others of the numerous hypotheses of mental and moral philosophy which the ingenuity of men have been fabricating since the world began; and of them it may be said, before the progressive, onward march of this science of facts, they are

> "Like the baseless fabric of a vision,
> Dissolving, leaving not a wreck behind."

The tender infant, when first ushered into the world, with the finest formed head, or the finest formed legs, is alike incapable of either thinking or walking, from the want of strength and consistency, or maturity and perfection of the members or functions by which these operations are performed.

It is only in the full maturity and perfection of our physical being that we are enabled to exhibit our highest or happiest intellectual efforts, or the greatest amount of physical energies.

The disease, decay, and gradual extinction of animal and intellectual powers appear the natural concomitants of extreme old age; this period presenting the melancholy spectacle of the tottering, imbecile, and weak old man—a perfect wreck both of body and mind.

The great poet and pet of nature so beautifully illustrates these different periods of man's existence, that we must beg to quote him.

> "At first the infant,
> Mewling and puking in the nurse's arms;
> And then the whining school-boy, with his satchel
> And shining morning face, creeping like snail
> Unwillingly to school; and then the lover,
> Sighing like furnace with a woful ballad
> Made to his mistress' eyebrow; then a soldier,
> Full of strange oaths, and bearded like the pard,
> Jealous in honor, sudden and quick in quarrel,
> Seeking the bubble reputation
> Even in the cannon's mouth; and then the justice,
> In fair, round belly, with good capon lined,
> With eyes severe and beard of formal cut,
> Full of wise saws and modern instances.
> And so he plays his part: the sixth age shifts
> Into the lean and slipper'd pantaloon,
> With spectacles on nose, and pouch on side,
> His youthful hose well-served, a world too wide
> For his shrunk shank, and his big manly voice
> Turning again toward childish treble pipes,
> And whistles in his sound. Last scene of all,
> That ends this strange, eventful history,
> Is second childishness, and mere oblivion—
> Sans teeth, sans eyes, sans taste, sans everything.

Concussions on the brain produce insensibility in a greater or less degree, proportionate to the injury sustained. The various degrees of intoxication are also productive of a corresponding degree of mental imbecility, amounting, in extreme cases, to total insensibility. This is clearly referable to the combined effects of the rush of deleterious gases to the brain, generated by the chemical action of this liquid fire, or alcohol, on the contents of the stomach, also from the repletion of all the vessels, particularly of the brain, thus producing a greater or less degree of inflammation or disease of this organ of the mind. Deleterious or noxious gases inhaled into the brain produce effects on the mind analogous to their character. The first stages of inebriation and the effects of "laughing gas" are very similar. Hypochondriacism, loss of children and lovers, or the frustration of any long-cherished passion, react frequently with a fearful energy on the whole animal economy, producing insanity, disease, and even death. Excessive mental agitation, intemperance. or excesses in any form, are therefore as prejudicial to the mind as body, and ought to be most sedulously avoided by those who wish to preserve their bodily and mental vigor unimpaired.

Cheerfulness and good temper are highly conducive to health; and happy are those who can preserve unruffled their equanimity under privations and disappointments. Grief, unmanly grief, ought to be beneath the dignity of proud, aspiring man: for

"Every grief but adds a nail to our coffin, there's no doubt,
Whilst every laugh so merry draws one out."

HARMONY OF SCRIPTURE WITH PHRENOLOGY.

The coming of the Messiah was a remarkable advent in the history of the world. Of him only it may be said, "He spake as never man spake;" and whose life, actions, and words are a true exemplification of Phrenology, and in precise accordance with its principles as a guide to human action. This affords at once one of the most conclusive evidences of the divinity of our Saviour, exhibiting in his God-like life those divine attributes of Christian charity, love, and forgiveness, which were so singularly contrasted with the bloody and barbarous Jews of that period, and indeed we may say of the whole world which, at that period of time, from the dreadful and exterminating wars which the most enlightened nations were waging against each other, had converted human beings into monsters, and this smiling world into one vast arena of blood and slaughter, wherein whole nations were oftentimes found exterminating each other, without regard to age, sex, or condition; thus surpassing in ferocity even the lowest orders of brute creation, who but rarely war with their own kind.

One of the most touching, affecting, and beautiful illustrations of character, found on record, ancient or modern, is that of the termination of the Saviour's sufferings. When under the excruciating pains of a violent and ignominious death, his divine exclamation was, "Father, forgive them, for they know not what they do!" Again; his golden rule, which comprises nearly all which a Christian or a philanthropist can require, viz., "Do unto others as ye would that others should do unto you," for thereby hangs all the law and the prophets. This beautiful exposition of moral conduct in man could only result in an entire knowledge of the capacities of man, and as peculiarly adapted to his state of being in that era, and also to future ages when man shall have acquired his highest degree of perfection, virtue, and happiness.

These evidences of the divinity of the Saviour proclaim aloud that "peace on earth and good will towards man" was the paramount object in view in his mission on earth. These beneficent views are singularly in accordance with the relative and social duties of man, as inculcated by the correct exposition of the phrenological doctrines when applied to the government of man in society, or in his individual capacity. This leads us to consider

"THE MORAL APPLICATION OF PHRENOLOGY."

In teaching the supremacy of the moral sentiments, it leads us invariably previous to action to ascertain "what is right," by an appeal to those monitors, Conscientiousness and Benevolence, which cannot lie, (although we believe various persons are differently constituted as to their acute perceptions of right and wrong, for instance as in the case of idiots, imbeciles, &c.;) we touch probably the most delicate ground on which the science stands; as by referring man's actions in a great measure to his organization, or as a consequence of his perfect or imperfect cerebral organization, or form of head, it has, we are aware, been considered by some most conscientious philanthropists and Christians as highly objectionable, as removing man's accountability for his actions to his Creator, and involving the dangerous doctrines of materialism and fatalism. But to the candid and unprejudiced observer it must be admitted we are dependent on our organization in some measure. We conceive this result is perfectly in accordance with divine revelation, which emphatically declares that to every man hath been given various degrees of talents—to one man was given one talent, to another ten, &c.—and that every man shall be judged according to his works, or the talents thus bestowed on him. Now does not this distinctly imply that human beings are differently constituted, and have as many and as various degrees of talent and temper as is compatible with the divine intentions of man's creation and his present existence?

Common observation strongly confirms this scriptural definition,

Inasmuch as it will be apparent to any who will investigate the matter that men are as differently endowed, both in their moral, physical, and intellectual character, as can well be conceived; inasmuch as we find some who would endure the most terrible privations, ay, even death itself probably, rather than to steal or murder; whilst we also do know that many unfortunates are the continual occupants of prisons, hospitals for the insane, &c., from either a defective organization, in a greater measure to be attributed to defective education, and partly possibly from the injudicious modes of punishments heretofore adopted for the avowed objects both of punishing and of prevention of crimes by example. Now, as to the infliction of unmerited punishments, we only indulge in the animal feelings in inflicting pain on others, and it is at best calculated only to irritate and harden the worst of criminals, and has a tendency to degrade the novice in crime to the level of the most abandoned, and lower him in the estimation of himself and others. As to the prevention, by the terrors of example, and punishment of crime in others, it is of a very doubtful expediency at the best; and regarded in the worst light, it outrages the moral sentiments of benevolence and conscientiousness, "in doing a positive evil, in the bare and very remote possibility that good may come of it." Hitherto it appears to have been more the object of men to invent punishments for crime, than to use any efforts in order to its preventton. This we conceive is the great error of all legislation without Phrenology.

It is hoped this digression will lead the seekers of truth from Phrenology to well weigh these considerations, as they involve questions of the highest import to the whole community, families and individuals.

The friends of the science are very sanguine in the hope, that great good will result from its general adoption in all public and private institutions, particularly of education, reformation, and mental culture, as by means of it great assistance may be rendered by its application as a powerful auxiliary in the noble cause of human elevation and improvement, both of national and individual character.

This is a science which perhaps has been more misunderstood and misrepresented than any other. Some have ignorantly, many wilfully, perverted its meaning and objects by levelling their shafts of ridicule and sarcasm, under the cognomen of "*bumpology*," &c Now it is true there may be occasionally bumps on the head, and there may be also depressions, but these are scarcely of any importance whatever in estimating the character; and when we say that many of the finest heads are entirely destitute of them, we shall perhaps astonish some whose heads may be highly embellished in this way, and who may suppose they have a strongly marked character, when probably they have one considera..y below par.

CHOICE IN MARRIAGE.

THE PLEASURES OF MATRIMONY AND WOMAN'S RIGHTS PHRENOLOGICALLY ANALYZED.

By adopting the science of Phrenology as the rule of moral conduct, we must be guided by a reference to the constitutional organization, and particularly the formation of the brain. It will be seen by a reference to its structure, (p. 89,) that the organ of Amativeness, or that portion which gives rise to the sexual feeling, alone occupies nearly one-third of the whole volume of the base of the brain, in addition to which the organs of Philoprogenitiveness, Adhesiveness, and Inhabitiveness are immediately above it, from which it may be inferred how large a portion of the brain is devoted to the social feelings, or those comprised in the domestic group.

In a phrenological analysis of character, we always assume the larger organs to control the smaller; consequently, the immense strength of these organs will be inferred, and their power on the character of an individual, for good or ill. In order to afford them every legitimate gratification, (for we cannot entirely suppress these feelings, neither ought we so to do,) we ought to be acquainted with their tendencies, their uses and abuses, and the mode of directing them, and this can only be properly effected by a close analysis of their various functions, as revealed to us by Phrenology, Physiology, &c.

By such an analysis we are led to the conviction that man is pre-eminently endowed a social being, or is wholly formed for society; also, that a state of solitude to him is a state of positive pain, precisely in proportion to the strength or development and volume of these organs, with the other portions of the brain. The relative proportions of these organs in the different sexes, with the differences in civilized and barbarous men, and the organs which control and direct them, may be seen by referring to the tables of admeasurements, as positive facts, which throw a flood of light on this subject.

It may there be seen how much larger, proportionably, are those organs which constitute spiritual love, as Veneration, Adhesiveness, &c., with love of offspring or children, in the female than in the male heads; whilst it may also be seen, that the simple feeling of desire, or animal love, (Amativeness,) is larger in the male—he being impelled more by love for A woman, than THE woman; whilst the exact converse of this is true of the female, which, added to her love of children, renders the marriage state so desirable to woman in every point of view, and absolutely necessary to her happiness.

Metaphysicians, who have argued the unlimited power of the soul over the body, or that man can begin and carry through a series of actions, independent of all cause and motive, we think were entirely ignorant of natural laws, as revealed by the operations of these organs on the animal economy, as we conceive they are an entire and complete refutation of such an opinion. In illustration of this view, we shall find the greatest and best of men have, at various times and in all ages, committed follies, and even crimes, under the strong impulsive power of this passion of love. These phenomena are perfectly incredible on the supposition of their actions being entirely independent of their organization. History furnishes so many singular facts, illustrative of the strength of this passion, in all ages and in all countries, that it has become a proverb, and it would here be work of supererogation to enumerate them. Sufficient to say, the very wisest and best of persons who ever lived have been, at some seasons, vanquished by this irresistible love.

One of the most instructive lessons for avoidance of this error, is probably afforded us in the history of those pious and holy, but mistaken men, who, imagining they were doing God service, have, under strong religious excitements, vowed eternal celibacy, in order to propitiate, as they erroneously imagine, the favor of Heaven, by outraging one of the best and holiest feelings implanted in our nature. The confessions of their tortures are almost appalling, and make us blush for the ignorance of poor human nature, and the misconceptions of duty and religion men fall into when they abandon nature for their guide. We can only regard such persons as monomaniacs, whose ardent love of religion, (as they interpret it,) has disordered their brain, or produced a morbid excitement of the moral organs, and who erroneously conceive the animal organs are their deadliest foes, because excesses have been committed when left unrestrained, and they think to repress them altogether. By so doing, they convert one of God's holy ordinances into a chimerical and most horrible phantom; and the impulse of this divine feeling they have tortured into the temptations of the Evil One himself, in resisting which they supposed they were "fighting the battles of the Lord," to win for themselves a glorious immortality.

How profoundly ignorant are such persons of the physical and organic laws which the truths of Phrenology are laying open to our view! This teaches us these organs of the propensities are among the holiest and best when directed by the intellect and sentiment; and in illustration of this fact, it will be found those nations who possess the highest or largest developments of the moral and intellectual faculties, are the most observant of their matrimonial obligations, and cherish these as the choicest boon from Heaven to man. Matrimony is also an institution of God himself, and by which woman is placed more nearly on a level with the boasted "lords of the creation;" for, amongst savages of all nations, she is treated as the slave of his passions and caprices, and from the degradation in which she is held by her brute tyrant-master, her whole being becomes deformed and deteriorated. Travellers frequently, in describing the brutal and hideous appearance of these savage men, sometimes inform us the women are still more repulsive in their appearance. [See Appendix.] How strikingly is this contrasted with woman here and in other highly civilized countries, where they are usually distinguished no less for their physical than their sentimental beauty. [See chapter on Physiognomy, also Admeasurements.]

This unquestionably results, in a great measure, from the superior respect and consideration shown them, which thus adds to their beauty and goodness, whereby all the superior sentiments are called into action. In this respect woman in this country may take precedence of all others, as woman here possesses greater freedom than elsewhere.

By a phrenological analysis of woman's character, she is not yet, we conceive, all she might be and is susceptible of realizing. We think she ought to be, in every respect, on a perfect equality with man, and every disability, political or otherwise, under which she labors, ought to be removed. Many employments, from which tyrant custom has excluded her, ought to be thrown open, and where nature or inclination points the way let her be free to exercise her gifts.

It is by the diffusion of knowledge, particularly Phrenology, the character of woman will be elevated and improved in the intellectual and physical organs. They are at present almost too much debarred from free air and exercise. Something might be learned from Physiology in this respect. We can conceive no reason why ladies should be entrusted with the very highest political power in Europe, and yet be excluded from many other subordinate trusts and employments. As the advocates of religion, temperance, morality, and virtue, they are, and would be still more, their efficient advocates and supporters.

But we are digressing. However men may differ in some respects on these points, all will be unanimous in the opinion that women are eminently qualified by nature to render the home of man literally "a little paradise" and home of the affections. How much do we not owe to her superior, soft, attractive graces, or higher sentimental character, [see Admeasurements,] in promoting the love of home, virtue and happiness! Indeed, no man can be said to possess a home, without a woman to grace and adorn it, and his happiness must necessarily be incomplete without her. Woman can be the only sincere friend of man in the hour of need. Man constitutionally cannot possibly feel that sympath for man which

88

A perfectly formed Female ad, with superior temperament.

HUMANITY.

See page 80. No. 1, Reptile, ascending to No. 4, Orang. This page
No. 5, Idiotic Human Animal, ascending to No. 8, Apollo

woman does, and no woman feels so deeply or so sensitively as the wife of his affections, who, united to him by every endearing tie of love, friendship, and family, sincerely sympathises in all his joys and sorrows. If in life there is one feeling superior to all others in extatic pleasure, it is when man turns from the cold, unfeeling world, to the bosom of his beloved one, and receives that sympathy of love he looks in vain for elsewhere, and however the storms of adversity may assail him without, he will find a haven and safe anchorage within his sacred home, lightening, if not removing all that mischance or ill fortune can huddle on his back. When thus cheered and invigorated by those he loves, he may bend, but will not break under the pressure of misfortune, and the sweet partner of his affections and happiness cheers him on his way, and participates in his weal and woe.

—————all the good we pray for in this life,
Is to be bless'd with one sweet, loving, fond, confiding wife.

In order fully to attain this blessing, one of the first requisites in a good wife is to ascertain that she has a good head, and all other good things will naturally follow, such as good temper, good health, good nature, good feeling, and, above all, good children, particularly if you yourself are also good. But, above all, in the choice of a wife, let sincere affection and real esteem be your guide. This will prove the greatest happiness of life, your lasting comfort, and a source of perpetual bliss.

Marriage, with love, is like a beauteous
Fountain of perpetual, never-ending sweets:
Without love, it is the source of hateful fear,
Of discord, strife, and jealousies without end.

As it would be impossible in our limited space here to do justice to this very interesting subject, we will refer the reader to our small pamphlet elucidating this subject, entitled "The Way to get Married; or Rules for a Happy Choice in Marriage," proving the necessity, happiness, and utility of marriage, founded on Phrenology and Physiology, natural and revealed religion, &c.

OF THE PHYSIOGNOMICAL CHARACTER OF MAN, AND OF THE DIFFERENT VARIETIES OF THE HUMAN FAMILY.

"And God said, Let us make man in our image, after our likeness. And God created man in his own image; in the image of God created he him, male and female created he them."

How ennobling to man this declaration of the Deity, by whom he is thus declared to be endowed with divine beauty. immeasurably beyond every other creature whatever. He was to resemble the Deity himself, the great and divine Author of all good.

Thus man, pure from the hands of his Maker, was endowed with a degree of heavenly beauty and intelligence corresponding to his state of innocence, happiness, and virtue. It will be interesting to show how far he has lost that heavenly type of his divine Creator, in a corresponding degree with his mental and physical degradation and debasement, as exhibited in his history in the various regions of the world. A brief outline of the most remarkable will be given in the *Appendix*, for the illustration of these facts in detail.

The cuts appended are intended, imperfectly, to exhibit the progressive scale of being, from the frog or reptile up to the classical Grecian profi.: of Apollo Belvidere, from Lavater.

To the love of nature, nothing offers a more delightful field for study and observation than an acquaintance with himself and his own peculiar organization and functions, as expressive of the various emotions, ideas, and sentiments of which we are susceptible.

Phrenology assigns the seat of every emotion and feeling as arising in the brain, or the cerebral structure.

Physiognomy, on the contrary, assigns the features as the origin of our various sensations, and to the countenance the manifestation of mind, disposition, talents, &c.

Now, on comparing the merits of these respective theories, we distinctly disclaim assigning to either of these instruments the origin of our sensations. We consider them as merely the instruments or mode by which we are enabled both to receive and to communicate our impressions of the external world; and in this view of the question, we might suppose the great Author of our being has created souls of different degrees of purity or loveliness, and assigned to each their appropriate habitations of beauty or deformity, both internal and external; for in our subsequent remarks (and common observation establishes the fact) there is a remarkable correspondence and harmony between not only the brain and face, but also in the whole of the organized structure, and we venture to say, this will be seen by all who are the least sensitive to beauty, (and who is not?) We sometimes meet with those delightful persons whom we are disposed almost to idolize at first sight, wholly from their external appearance or physiognomical expression, whilst we also experience strong feelings of aversion to others, simply from unfavorable appearances. We doubt not but every man's countenance is the index to his character, (not the cause of it,) were we but endowed with sufficient powers o' discernment to read it. We have ourselves invariably observed hat where the destructive propensities are very small, it is next to impossible for persons thus constituted, strongly to depict these passions by the countenance. So of benevolence, &c. We also believe there are many exceptions, or rather that many most noble natures are overthrown and prostrated by the force of untoward circumstances, (and so misunderstood,) and the milk of human kindness turned to gall and bitterness.

As a further illustration of the truth of Physiognomy, let us compare the dog-like skull of the cannibal, (page 45,) and then read the accompanying physiognomical description, and judge how nearly it accords with the horribly repulsive character given of these worse than savages, by various travellers.

To such descriptions it will only be necessary as a contrast, to conjure to the mind's eye those forms of beauty and loveliness with which he is most familiar and enamored; or let him turn, in this volume, to some of the imperfectly drawn sketches of American beauties, Shakespeare, &c., and he must admit there is just as great and striking a dissimilarity in their phrenological developments and physiognomical expression and appearance, as in their very widely dissimilar and striking characteristics. [See tables of measures.]

For our purpose it is not necessary to go into minute detail on this subject, but let these strongly marked and diversified facts speak for themselves, as there exhibited. This leads us to consider

THE ADVANTAGES OF A PHRENOLOGICAL AND PHYSIOLOGICAL STUDY OF NATURE.

To a contemplative mind nothing offers so much to admire as the physical organization and mechanism of man. We may well exclaim with the Psalmist of old, "Lord, I am fearfully and wonderfully made." Nothing, probably tends to awaken a sense of our dependence on the divine Author of our being, more than an investigation of our own organization, structure, and bodily functions. Indeed, we think this as a branch of study ought to be paramount to most others. More particularly is it of the highest importance to the mothers of our country. We would ask, what earthly blessing can equal that inestimable feeling of happiness we experience in the enjoyment of a vigorous state of health. Indeed, without this blessing, the greatest of earthly enjoyments can avail us but little. Now without a knowledge of some of the most important functions of the human structure, how can we hope to treat ourselves or children in a manner conducive to health?* Let me ask, would we not require a machinist who undertook the charge of a steam engine, to understand its mode of operation, of shutting off and putting on steam at appropriate times, and conducting every other department of duty

* Dr. Andrew Combe on the Physiology of Digestion, ought to form a part of every family library.

with correctness, fidelity and despatch—a knowledge of which can only be acquired by a long and studied acquaintance with the operations of the machine and its various parts. What would be the results of placing one in charge of the engine who had never received the least instruction upon this subject? Why, we should apprehend the most disastrous consequences to ensue.

And yet, I should say, precisely in this condition are those parents placed, when the charge of a family devolves on them without their having the least previous acquaintance with the structure and organization of the tender and delicate beings thus entrusted to their charge. It is very true that the instinctive love and fondness we have for children might prevent us committing any very glaring mistakes; but although nature has endowed the whole of created beings with this necessary faculty in a corresponding ratio with the care and attention necessary for their health and preservation, yet to man has she denied those strong instinctive propensities possessed in so remarkable a manner by the inferior order of animals for the care of their young, &c. Yet as man, of all other beings, comes into the world more helpless, more tender and delicate, requiring also much more varied nutriment, and for a much greater length of time, but as she has not been thus liberal to him, she has far more than compensated in this respect by the endowment of that God-like attribute which we call reason, and by the cultivation of which we can render ourselves infinitely more competent to discharge our duties. By judiciously exercising his reasoning faculties, man can immeasurably surpass the brute creation, in this respect as in every other, as instanced by his skill in the management of animals, &c. But how does the case stand at present, with regard to his own offspring? Why, it has been clearly ascertained by statistical tables that in Europe an immense portion, computed to be one-third of the whole number born, perish in their infancy. And can we arraign the wisdom or goodness of Divine Providence in this fatality of the human species in the earlier stages of existence? By comparing this frightful mortality with that of the lower order of animals, we have indeed reason to blush at the comparison! No such mortality exists amongst the inferior animals, or by any means approaching it, and we must come to the inevitable conclusion that in spite of the boasted improvements of the age, we are yet lamentably deficient in this most important matter. Every person having the happiness and well-being of others in view, ought to direct his most earnest attention to this subject, in order to avert the fatal consequences here alluded to.

Mr. G. Combe [see Appendix,] also has acquired a very great and deserved celebrity from his beautiful exposition and philosophical mode of investigating and applying the science as a moral rule of action. He contends, and beautifully exemplifies, that by studying and adopting the rules which regulate physical nature, as divulged by the studies of Phrenology and Physiology, we may attain the highest degree of health and happiness; and also that by studying the laws of health, we may infinitely improve ourselves both in our physical and moral being, and also that these conditions are much more nearly allied than is generally supposed by superficial observers

It will be found there is nothing more conducive to happiness than health; and yet it is astonishing how much this condition is neglected, both in regard to the laws of hereditary descent, and also those of pure air, exercise, and dietetics. How many of even the better educated act as though there were no physical laws, which govern existences, and of animated nature, of which we form a part. How profoundly ignorant are many on these subjects! And yet, without health we are unfit to exercise our intellectual faculties, to receive education, or to fulfil our duties in our social relations. Without health, we may become a burthen to ourselves and others; and yet we are continually outraging some of the most obvious and most salutary laws, which are unquestionably framed for our good, and by a moderate attention to and observance of which, we may immeasurably promote our own and others' happiness.

Could man but implicitly submit himself to the will of his Creator, or, in other words to his natural laws, and their peculiar application to himself, as revealed by a thorough and accurate acquaintance with his own innate feelings and fundamental, primitive character, as developed by Phrenology, he would infinitely advance in religion, virtue, and happiness. This naturally leads to the vast importance of the

HEREDITARY TRANSMISSION OF ACQUIRED HABITS AND IMPROVED ORGANIZATION.

In showing the remarkable phenomena exhibited by nature in the transmission of hereditary qualities, we shall be greatly assisted in our researches after truth by a comparison of animals, particularly those which are under the guardianship and tutelage of man, as the horse, dog, &c. Of the infinite variety of these animals, we must regard them as having each sprung from a common stock; and the great diversity of talents, temper and disposition of the dog are clearly traceable to the care and attention bestowed on him by man, not only in training and teaching them particular habits and acquirements, but also in the very great care and expense incurred by individuals in selecting the most beautifully formed and more perfect specimens to perpetuate their respective forms and qualities. So great has been the success attending this plan in England, where it has certainly been carried to a wonderful degree of perfection, that it is found possible to produce dogs, horses, &c. of almost any required qualities, simply by a judicious selection of the parent stock.* To prove the possibility of this transmission of qualities, it is well known that the more perfectly educated dogs, as the pointer, setter, &c., not only themselves but also their offspring, inherit in a remarkable degree those rare qualities for which they are so much prized.

Now, in order fully to understand this subject, we must bear in mind that the qualities here alluded to are by no means their natural propensities, but are all acquired, and show how much nature may be moulded and even entirely changed by art. The same may also be said of the human species.

* Dog fanciers, &c., are frequently much better acquainted with the pedigree and ancestry of their horses and dogs, than of their own families and appear to think it of greater importance.

Domestic (or Improved) Dog, with Benevolence, &c

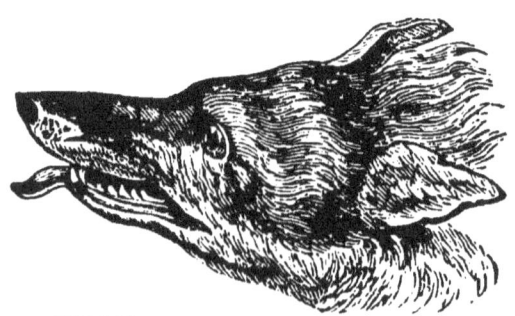

Wild Dog or Wolf—no Benevolence, &c.

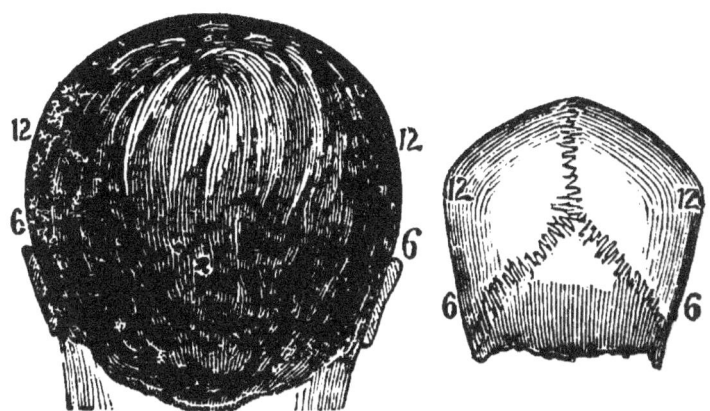

REMARKS ON THE MURDERER R——.

The drawing of the large head above is of a most cool and deliberate murderer, whose name we will suppress in consideration of his numerous relatives. He had received a good education, and for some time was captain of a steamer plying to New Orleans He subsequently engaged in the Texian service, and held the rank of lieutenant at the period of his committing the murders of two individuals, (a medical gentleman and a passenger,) who had left the Texian vessel in the Gulf of Mexico, near the mouth of Bayou la Fourche. It appeared in evidence, on the trial, that R—— and an accomplice left the vessel on the pretext of gunning, when in reality they pursued the two gentlemen up the Bayou, each agreeing to participate in the murder, but the courage of his accomplice failing, this Lieut. R—— shot both the unfortunate gentlemen. They then sank one body, by tying heavy shot to it, which they had brought from the vessel for this purpose It was their intention to sink both the bodies, but they found the shot would not be sufficiently heavy, so they dragged one on shore, and buried him in the swamp, where he was afterwards found. The motive for the deed could not be ascertained, as neither of the gentlemen alluded to had property of value with them. These criminals had the audacity to take back some of the murdered victims' articles on board the vessel, and as they had been absent all day, and were seen to secrete some articles under the spars on deck, it led to some suspicion, and, on examination, proving to be articles of the two gentlemen who left in the morning, they were ironed whilst in their cots. taken to Thibadeauville, and both found guilty. Johnson. the accomplice, was repriev ed; this R—— was executed, and under those appalling circumstances, as he had himself asserted, and as the extraordinary head would indicate, he retained his composure and self-possession in a most extraordinary degree. He appeared the least affected of any present. His conduct strongly contrasted with the sobs and lamentations of the females, (who could see the execution from their dwellings.) All else was as quiet as the grave. The narrator has seen eight poor malefactors suspended at once in the midst of a populous city, without experiencing sensations of horror as at this time. The sun was shining in his meridian splendor, with a glorious brilliancy. Enormous trees, with their giant arms festooned with long, sweeping, but sombre looking Spanish moss, filled the background, conveying a melancholy but picturesque effect to this scene of extremest human suffering and anguish, and, it may be added, of almost super-human courage exhibited, worthy a more glorious field.

At his own request, he had been furnished with an extra length and thickness of rope; the gibbet was of an extra height, that he might have five feet to fall, in order, as he said, to make quick work of it. He required no assistance in ascending the ladder, or in placing the rope round his neck, nor was his face covered. He thus stood firm as a statue for several minutes after every thing was ready. Bowing gracefully to the few assembled, he gave the preconcerted signal, the sheriff withdrew the bolt, and, with a tremendous crash and dislocation of the neck, in a few moments, without a struggle, a distorted, hideous, livid corse swung before us, a terrible example to evil-doers.

His head, it will be seen in the measures, is strikingly like Deaf Burke, the noted English pugilist, who has, we believe, killed several in boxing. It is also a remarkable fact, that in his fight with the Irish champion, O'Rourke, about that time, he nearly killed him in two or three rounds, and had not the deaf one been furnished by an American gentleman present, with one of those huge cleavers called a bowie knife, with which he fought his way out of a mob, he would have been destroyed on the spot by the infuriated Irishmen and friends of O'Rourke, who was the boxing champion of New Orleans.

It is also said of this murderer that he performed a similar exploit in New Orleans at a former period.

When the author visited New Orleans jail, he found no difficulty in selecting the worst criminal there. He was a most notorious, pugnacious person, and his fellow prisoners dubbed him John Bull. His head was very similar to this murderer, but he would not be measured, although the author offered him a sum, as he imagined and (like many other criminals,) was fearful some actions of his past life would be revealed by such an examination.

NATIVE HINDOO SKULL, FROM CALCUTTA.

The small skull is that of a Hindoo, being one of six which were procured from Hindoostan by Dr. J. V. C. Smith, of Boston. The heads of each are strikingly similar, as the tables of measures show, being very small in Destructiveness, or 6, and very large in 12, or Caution. A drawing of one suffices for the whole. They are all strikingly illustrative of their national defect, timidity and cowardice—so much so, they have been the prey of every invader, from the earliest records to the present time. Hindoostan has been conquered and successively held in bondage by the ancient Greeks, the Persians, the Tartars, the Moguls, and more recently by the British, who, with a comparatively small force, still hold this country in a state of vassalage, and exercise a despotic jurisdiction over the natives, chiefly by fomenting petty jealousies amongst the native princes, setting them at war with each other, and then, under the pretence of arbitrating their differences, levying enormous annual contributions from each, who, in their turn, levy these on their people, producing in some seasons great destitution and misery; thus perpetuating one vast system of plunder, unworthy a great and enlightened people.*

When a youth, the author visited Calcutta, when the Burmese had obtained some advantage over the British, and it was thought necessary by the British East India government to obtain an additional volunteer marine corps from the British ships then in port at

---

* The British tory government, under the old idiot George III., would have succeeded in affording just such protection to the Americans, had the heads of the Revolutionary patriots been like these. That monarch's head, as exhibited on the bronze equestrian statue in London, the author has phrenologically examined, and pronounces it but a very little superior to the Idiot, (see portrait No. 2,) although the work of the first sculptor of the age. The whole head is the most repulsive he has ever seen. A phrenologist almost shudders to look at such an inferior head.

This idiotic monarch has also generally been described by Byron, and other independent writers, as having a dog forehead. Of course, it is a matter of history that he was confined as a lunatic for many years. The portraits we have of him are grossly flattered, as well as those generally of Queen Victoria, and others of royal blood, both male and female. The descendants of this old fool are, and have been in their private lives, with scarcely any exception, the most notoriously vicious characters in Great Britain. They have expended millions on millions of the people's money in every species of crime and debauchery, of which the world scarcely affords a parallel, some of which are almost unknown here.

It was the author's first object, on arriving in this country, to forswear allegiance to such superlative villanies, and he cannot here help recording his opinions and his detestation of their tyranny and robberies on their own people. It is hoped the English and American people will unite against their only common enemy the aristocracy of Great Britain.

(See last page.)

Calcutta, the Sir Thomas Grenville's crew and officers, about one hundred, volunteered. The government, in place of these, furnished three hundred of the most robust Lascars to stow the ship's cargo, &c., and at the end of two weeks they had scarcely accomplished as much as one hundred Englishmen would do in half that period. In lifting any article, however light, as many would cluster around it as nearly to hide the object. In forging a small piece of iron, four or five will station themselves around the anvil, striking at intervals. He has also seen four using one large saw, and so on with their other tools. They have smaller heads than probably any other people whatever. They appear to have remained stationary for ages.

A PLAN

WHEREBY ALL PERSONS MAY BECOME THEIR OWN PHRENOLOGIST; ALSO TO CHOOSE A WIFE, COMPANION, OR FRIEND.

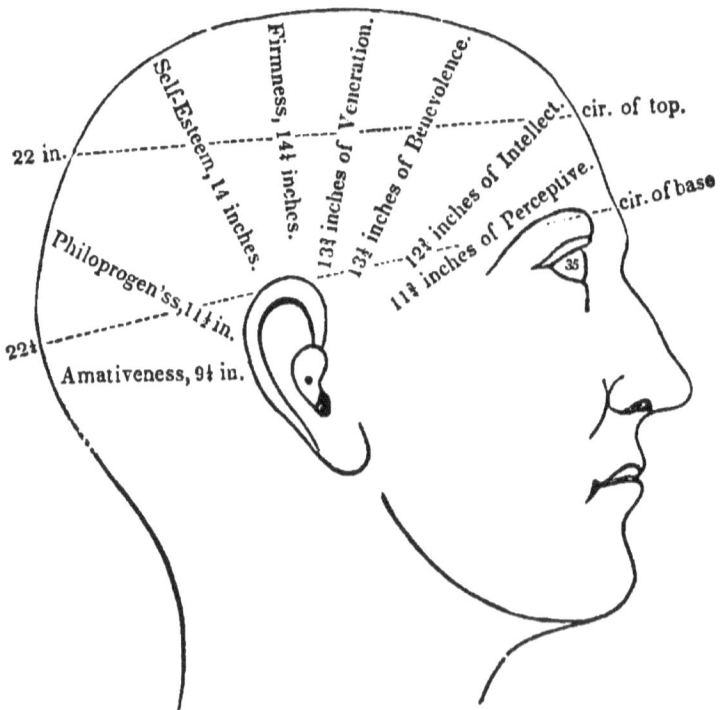

Apply the end of a common tape measure exactly to the meatus or opening of either ear, and draw it round horizontal, (the head being perpendicular,) to the opposite ear, and the number of inches gives the strength or size of the particular organs it covers. The same mode may be resorted to in the direction of the dotted lines, to ascertain the proportion of the other organs so specified.

The whole may be compared with the sizes here given of F. Coombs's approved bust, (but which, owing to mistake, is rather too large in Amativeness.)

This mode will afford a very simple and very correct way of ascertaining the leading phrenological character.

Superior female heads are usually rather smaller than the male head which these admeasurements represent.

The measure of the base of circumference of head will contrast with the portion immediately above it, (the dotted lines,) and serves to show the controling organs, Intellect, Caution, &c., as opposed to the propelling or animal powers, or base of brain. The measure from bottom of forehead, or Individuality, to the occipital spine, (or small, bony projection near the base of skull behind,) serves to show the intellectual and sentimental powers, and when larger here than in the bust is a very favorable indication of character.

By this method, we only obtain the general outline of character, but it clearly shows that which is most important to know, viz., the proportions which the various parts of the brain or implied leading characteristics bear to each other. This admeasurement of the head also embraces, besides those named on the cut, the organs which the tape passes over, for which reason we may take it as a general classification of the various organs, as they are embraced in groups by this measure, nearly all being of a similar character, tnrough the line which the tape covers, and the measurements in each region give a very fair estimate of the general character, much more to be depended on than the mode of guessing by some phrenologists, who seldom resort to measuring the head, and who must consequently be more liable to error of judgment.

This mode of estimating character has been too much overlooked, and to promote this object, the tables of measures are furnished in this book, which he means to adopt as his standard, also for the purpose of eliciting truth by such measures. As an instance, should any gentleman remarkable for his philanthropy be found to measure very low in the region of Benevolence, we will abandon Phrenology, and proclaim it to be founded in error and misapprehension. But we have not the slightest fear of such a result. The tables of admeasurements show most conclusively the general characteristics of the various individuals. He will still resort to the same method, and intends to collect as many admeasurements of remarkable persons as he possibly can, for which purpose he invites all public or remarkable personages to have their heads measured, free of expense, for future publication. The admeasurements of very remarkable, pious, amiable, or exemplary ladies are also much wanted as a contrast to inferior heads.

The author extremely regrets that an admeasurement of his own head, taken by the same instrument six years ago, was lost at the printer's, as it would have conclusively shown a very considerable increase of size in that period, without an increase of personal size or weight, clearly showing that cerebral action, as in the exercise of his profession, evidently tends to enlarge the brain.

He would here take this opportunity of qualifying an expression made in a former part of this book, as to the sizes of the intellectual organs of savages, &c., as one or two of the North Western Indian Chiefs appear to be an exception to the rule of absolute size; but then it will be seen, relatively to the other parts or base of the brain, they can scarcely be called large. Those chiefs were gigantic men, most of them weighing over two hundred pounds, and the most athletic men he has seen on the continent of America. The author had an opportunity of hearing Keokuck, the head chief, make a set speech, of nearly one hour long, before a grand council of Indians. His eloquence was the most impassioned and energetic he ever heard, and the action chaste and perfectly appropriate. The language of the Indian is exceedingly guttural, and in perfect unison with the strong, warlike passions which they so delight in expressing in their war-whoop, words, and actions. Their language is peculiarly harsh and discordant, and appears almost incapable of expressing the softer emotions. Keokuck's speech will ever remain fixed in memory, as an example of the stern, the fierce, the terrible. He has never seen it surpassed by E. Forrest himself. It was equally picturesque, and standing as he did the ardent champion of a remnant of his race, invested him with a romantic interest which will not be readily forgotten.

The author has visited amongst the North Western Indians, and also, a few years since, the Hindoos of Asia, and in boyish sport has chased many of the latter; but a whole regiment of soldiers will scarcely suffice to make the former run. They live entirely on animal food, if they can procure it. He has seen them eat flesh of any description, scarcely half sodden, which they prefer, without bread or vegetable of any kind. The females are not allowed to partake till the men have eaten. They may be said to be perfectly carniverous. (See their measures.) Also the East Indians, who, on the contrary, are remarkably abstemious in their diet, will scarcely destroy animals for food, but subsist chiefly on rice, vegetables, &c. Indeed, so tender are they of animal life, they have hospitals provided for many of them, and persons injuring them are compelled by law to support them and pay the doctor's fees.

APPENDIX,

ILLUSTRATIVE OF PHYSIOGNOMY

[Referred to at page 90.]

Bleumanbach divides the human family into five different varieties, but which perhaps may be more properly called races. In their physiognomical and other peculiarities he thus classifies and describes:

1. The Caucassian race, inhabiting the greater portion of Europe, including the United States of America and some portion of Asia, are distinguished by a beautifully soft, fair, and transparent skin, susceptible of every variety of tint; the hair fine, long and curling, of various colors. The skull is large and oval, and its anterior or intellectual region full and elevated. The face is proportionably small, of an oval form, and delicately formed features. The chin full and rounded, and the teeth nearly perpendicular. Portions of this race early attained the highest degree of civilization and intellectual attainments.

2. The Mongolian race, inhabiting central and eastern Asia and some portion of the North American continent, are characterized by a sallow or olive skin, straight hair and thin beard; the nose broad and short; the eyes small, black, and oblique; the lips curled; the cheek-bones broad and flat. The skull is oblong-oval, flattened at the sides, with a very low forehead. In their intellectual character they exhibit considerable talents and are somewhat susceptible of education and improvement.

3. The Malay race are inhabitants of the Polynesian group of islands; also Madagascar, Borneo, &c.

This race are characterized by a dark complexion, varying from a tawny hue to a very dark brown. Their hair is black, coarse, and lank, and their eyelids drawn obliquely upwards at the outer angle. The mouth and lips are large; the nose is short and broad, and apparently broken at its roots. The face is flat and expanded, the upper jaw projecting, the teeth salient. The skull is high and rounded, and the forehead low and broad. This race exhibit great difference and varieties amongst them. Some are active and ingenious, but very degraded on the whole, and are nearly the lowest in the scale of humanity.

9*

The head of the Malay is large, and the nose short, depressed, and flat at the nostrils. The eyes are small, black, oblique, and expressive. The face is broad and very prominent, and the mouth and lips are large. Their color is a brown, with a bronze tint.

The skull of the Malay is very low in the forehead; the cheek-bones are high and expanded; the jaws are greatly projected, and the upper jaw, with the teeth, is much inclined outwards, and nearly horizontal. The teeth remarkably fine and strong.

A portion of this family, living on the Island of Battas, are the most remorseless and habitual cannibals on the face of the earth. Nay more, they not only eat their victims, but eat them alive, or do not previously put them to death, and these are not unfrequently their own people and relations. Prisoners of war are eaten at once, and the slain devoured without cooking.

In the great Island of Borneo, the Malays have possession of the entire coast, and the mountainous region of the interior is peopled by the savage Dayacks and Eidahous, who are represented as being fairer than the Malays, but still more sanguinary and ferocious.

The Polynesian family, a portion of whom, the New Zealanders, are the most sanguinary and intractable. Their combined treachery, cruelty, and cannibalism have made them proverbial since the discovery of their island. Capt. Crozet, whose crew they attempted to destroy, thus describes them:—" They treated us (says he) with every show of friendship for thirty-three days, with the intention of eating us on the thirty-fourth."

The Fegee Islanders vie with the New Zealanders in treachery and cannibalism. Capt. Dillon gives a heart-rending account of the murder of fourteen of his men, who were subsequently baked in ovens, and afterwards devoured in his presence.

4. The American race is marked by a brown complexion, long, black, lank hair, and deficient beard. The eyes are black and deep-set, the brow low, the cheek-bones high, the nose large and aquiline, the mouth large, and the lips tumid and compressed. The skull is small, wide between the parietal bones, prominent at the vertex. Their characteristics are slowness and distaste for acquiring knowledge, stern, unyielding, proud, revengeful and cruel. [See table of admeasurements of distinguished chiefs, who are mostly gigantic men, and have unusually fine heads for savages, far the best in the collection.]

5. The Ethiopian variety, inhabiting the greater portion of Africa and the continent (it may be called) of New Holland, and islands adjacent, are characterized by black complexion, black, woolly hair, eyes large and prominent, the nose broad and flat, the lips thick, and the mouth wide. The head is long and narrow, and the forehead low; the cheek-bones prominent, the jaws projecting, and the chin small. The several nations comprised in this variety are rather diversified, but extremely low, and amongst the most degraded of the human species; amongst whom are the Hottentots, whose complexion is of a yellow brown, or bright olive; the hair black and woolly, and very small beard; the backhead large; receding forehead, and wide, large face; the eyes far apart, the nose broad and flat; and the women if possible more repulsive than the men. They are inveterately indolent and gluttonous, devouring every kind of

animal refuse that falls in their way, without preparation, and then throw themselves down and sleep off the effects. Their dwellings are mud-hovels, bushes, caves, or clefts in the rocks. Many go naked, without shame; others partially cover themselves with the skins of animals they kill. They have no more notion of decency or cleanliness than animals, are robbers by profession, and kill everything indiscriminately which they cannot carry with them.

The Bosjesmans are far more degraded and savage than any other Hottentot tribes. Some maintain that they are different from the Hottentots, and constitute the ultimate link in the scale of humanity.

The face of the New Hollander is ugly in the extreme; projects greatly from the head, and the mouth is particularly prominent, with very thick and protuberant lips. The nose is flat and broad and the nostrils expanded. A deep sinus separates the nose from the forehead, and the frontal ridges overhang the eyes, while the forehead is low and slopes rapidly to the top of the head. The hair is often long, very coarse, and frizzled, yet rarely woolly. They are perpetually engaged in war, and seldom if ever pardon an enemy, but generally kill and eat them. Their courtship, if such it can be called, consists in violence, and their women are treated through life with unparalleled brutality. They are to the last degree filthy and gluttonous in the extreme.

A portion of the Australian family, inhabiting the Island Andaman, are of small stature, slender limbs, protuberant abdomen, high shoulders, and large faces, exhibiting a horrid mixture of famine and ferocity.

Foster compares the people of Malicolo to monkeys, and asserts that he had seen no negroes in whom the forehead was so depressed.

This family is also found in the numerous islands near Guinea, New Britain, Admiralty, and Hermit Island, &c.

The origin of color, and other differences amongst the human family, have frequently employed the pens of historians and philosophers. Here is one opinion attempting to elucidate this subject, by Dr. Caldwell:

"It is computed, by the Mosaic account, that about four thousand one hundred and ninety years from the creation, Noah and his family left the ark, who, from the most undoubted authority, were of the Caucassian or white race; and yet we have the most abundant historical evidences, that above three thousand years ago the Ethiopian or negro family were known as inhabiting a portion of Africa, and possessing their present characteristics. Consequently, if of the same race as the white, the change of color, features, &c., must have been effected in less than one thousand years, and then remained stationary to the present day."

To illustrate the fact that color is not the effect of a warm climate or exposure to the sun, nor the peculiar characteristic of the negroes alone, some of the Esquimaux (eaters of raw flesh,) on the icy shores of the island of Greenland, are extremely dark and many of them at Oppemwick are quite as dark as the mullatoes. Crantz, the missionary, says that they are crafty, sensual, ungrateful, obstinate, and unfeeling. They also devour the most disgusting animals, uncleaned and uncooked. Their mental faculties present a continued childhood. They are fickle and facetious, and their connubial infidelity is a proverb among voyagers. In gluttony, selfishness, and ingratitude, they are perhaps unequalled by any other people in the world.

[Referred to at page 93.]

The Diary of Mr. Combe, in his late tour in the United States, is somewhat more impartial than any of the preceding English tourists, yet does not, we humbly conceive, fully appreciate the great differences in the monarchical and republican forms of government on the character of man. There appears, though to a less extent, the same carping at unimportant trifles, and a want of comprehensiveness of the magnificent whole, as presented by the entire American people, more particularly as evidenced by the gigantic strides, within a very few years, in every useful art and science, and probably in none so much as those immense public works of utility, the railroads and canals, which are intersecting every part of the Union, and uniting all the people in one community of purpose and feeling. In a military point of view, also, these are of incalculable importance, as by this means of communication an army of one hundred thousand men can be concentrated in any one point on the Atlantic north of the Potomac in twenty-four hours, rendering us invulnerable to any power which can be brought against the republic by sea or land. The Americans may proudly contrast with the same race of people under a monarchical form of government, but let it be a loving, a generous comparison.

We must also beg leave entirely to dissent from Mr. C., in his views as to the justice and expediency of the movements of the chartists in England, who are contending for vote by ballot and universal suffrage, and whom we conceive to be engaged in the most righteous cause that men were ever yet employed in. We insist upon it, nothing but lunacy, idiocy, or crimes can justify withholding from, or depriving any man or body of men, the privilege of managing their own affairs. Shall we be told by Mr. C., or any other gentleman, that because some disorders may possibly ensue, (and this is merely his gratuitous supposition,) that we are still to continue the same villanous state of things which has reduced the true nobility of England and Ireland, (the working people,) to the condition of the poorest slaves in existence, (excepting in the actual sale of their bodies ?) And still, according to him, we are to persevere in this same course, which has so brutalized the people that he attempts to show they are unfit to exercise the privileges of men. Away with such unjustifiable legislation. Such arguments, no doubt, have their weight with the contemptible aristocracy of Great Britain, who really seem to imagine that all other people were born like beasts to bear their burthens. By a strange perversity of reasoning, totally unexpected from Mr. C., their tyranny and oppression are to be continued until these same causes which have produced this state of things have resulted in a state of things precisely dissimilar and diametrically opposite !

If we comprehend Mr. C., he deliberately advances this as his opinion, which we most solemnly protest against. Notwithstanding our admiration for him on other subjects, on this we are enlisted body and soul, and will not give up our opinion to any man living.

By the same parity of reasoning, the kind and generous slave-holder thinks he is doing good service in retaining his slave in his obedience until he has qualified him for what he conceives to be necessary for his happiness and comfort, and is competent to enjoy and make a rational use of his liberty Abolitionists will tell you how far distant such a day will be with those whose optics are so blinded by prejudice, they can only see the ignorance and debasement of the slave, and not the causes which have led to it, or the unalienable rights of man to be free. Both these systems of slavery in England and America equally tend to brutalize man. And those who hold them in bondage have the same kind charity for them—to take care of them until they are fit, in their estimation, to receive their freedom.

Such are the robberies and exactions on the poor working citizens of Great Britain, that this class have rarely the means to purchase books of popular authors, from the price being usually three or fourfold the price the same books are sold for in this country. Hence the press and nearly the whole of English literature is and must be purely aristocratic.

Mr. C. also complains of us in this respect, as doing a great injustice to authors; but Mr. C. might equally well complain of the whole world besides, for all freely avail themselves of English, French, and German literature, without this complaint being brought against them, and we see no injustice in our availing ourselves of the same privilege on a larger, which others do on a smaller scale; for we see no reasons to make distinctions of persons. If English authors appeal to the generosity of the American people for some remuneration for their works, we think it would be conceded; but we cannot conceive why they should levy contributions, on this score, upon us more than any other people whatever, from whom we never heard of their making any such demands as from us.

That the Americans and the whole civilized world are deeply, immeasurably indebted to English literature, must be conceded; whilst it may also be complained of that the spirit of English literature is decidedly aristocratic and extremely injurious, as it so largely contributes to the formation of character; and most of the vices of aristocracy in this country are the injurious effects of English literature on public taste, which is and must be decidedly aristocratic until a new order of things appear in England, which we hope is not far distant, when the sovereign people will sweep their barbarous oppressors like chaff before the wind, by the force of public opinion, when once they are thoroughly aroused to their own rights by the agitations, &c., of the chartists. America, we hope, is destined largely to contribute to this change. Then will the English and Americans be united as one people in a noble and generous rivalry of love; but this can never be until the whole people of England have assumed or possess a share in conducting the affairs of the government, and have driven their present tyrants and enemies of all mankind from the power which they have so unjustly usurped, and made themselves odious in the sight of all mankind, from India to the poles.

Mr. C. relates an amusing incident occurring on board the steamboat, as he was proceeding to Albany. It appears a passenger had unwittingly supposed that, as Mr. Combe was a phrenologist, he also examined heads. But Mr. C. informs us he does not in public, or for pay; which we think is to be regretted.

On the same page, (54, vol. 2,) Mr. C. does an injustice to a most worthy, sensitive young gentleman of Kentucky, (Mr. Porter,) whom Mr. C. says we made a show of, to add to the attractions of Phrenology. Now, the fact is, Mr. P., (the Kentucky Giant, as he is usually called, being the tallest man in the world,) had actually refused ten thousand dollars a year, if he would consent to be exhibited, but he declined. [See his Self-Esteem.] Major Stevens, also alluded to, was engaged as secretary. Mr. C. examined their very singular and most extraordinary heads in public, and lectured before twenty thousand persons in New York, in five weeks, of course owing to this attraction. [N. B. Wanted just such another couple of extraordinary heads to lecture from. See their admeasurements.]

Mr. C. is not the only person to whom ludicrous mistakes have occurred from similarity of names. Whilst the author was at Saratoga, an unfortunate tailor of Albany sent a bill and pair of pants, (not G. M.,) to Congress Hall, addressed "Mr. Combs, Phrenologist," whilst he was at that house. This was rather ludicrous, as they were a complete misfit, and the author never ordered or purchased a pair in the city of Albany in his *itinerising*.

As to the similarity of names, F. Coombs will be here excused for mentioning that his family are, and have been for fifty years past, extensively engaged as architects and builders, 9 and 10 Benett's Hill, St. Paul's, London. One of his relatives also is the architect of the most costly public edifices now erecting in the United States.

F. Coombs having had twenty years' experience in Europe, eight years in the United States and Canada, and several months' residence in the East Indies, hopes at a future day to give his impressions of the Anglo-Saxon family, as elicited under the governments republican, monarchical, and colonial. He had, previous to his last visit to England, foresworn allegiance to the British crown. America is now his home, and although born within the very shadow of St. Paul's, London, he shall call himself a Yankee phrenologist. He detests monarchical abuses and robberies, and none so much as the British tories, whom he conceives the greatest and most impudent robbers on this earth, not only of their own victimised, impoverished people, but of all creation, wherever they have the power; as witness their robberies in India, China, &c. But thank God, this they could not do for 'Jonathan' when in his infancy of growth, and he has nothing to fear now from all the tories in the world.

We are tempted to preach a crusade against these human harpies, or thieving tories, who are the curse of this earth. They have impoverished and nearly ruined dear old England, and reduced her noble sons and daughters to the condition of the veriest slaves on earth. Great God! shall such things always be?

Some of these hireling tourists are their strongest allies and supporters, by misrepresenting everything here, generous, free, or American. We certainly feel induced to publish also our tour, or peregrinations, to show how much the Americans and their free institutions have been belied. The author has experienced the kindness of the Americans, from the President in his palace to the squatter in his log cabin on the banks of the Mississippi, and strongly and utterly condemns the impressions of most preceding tourists, as being infinitely different and totally at variance from the truths he has arrived at, after a residence of above seven years in this country.

AMERICANS, READ THIS!

PHRENOLOGY *versus* ALEX.R McLEOD.

THE EXACT AND EXTRAORDINARY MEASUREMENTS

OF THE

Notorious Loyalist, Alexander McLeod,

WHO WAS TRIED FOR MURDER AT UTICA, ALSO THAT OF A

MURDERER,

LIEUT. R., OF THE TEXAN NAVY,

EXECUTED IN LOUISIANA IN 1837,

CONTRASTED WITH THE ACTUAL MEASUREMENTS OF

The Rev. W. E. Channing, the Philanthropist of Boston,

AND SOME OTHERS, INCLUDING

Wm. Lloyd Mackenzie, the Patriot ex-Mayor of Toronto

ALSO OF

E. A. Theller, Brigadier General

In the Canadian Republican Service; both of whom have lost all their property and barely escaped with life, for advocating Liberty and opposing the odious British Tories in Canada.

The above Heads have been each very carefully measured and examined by F. Coombs, practical Phrenologist, who submits this test of the Science to the American People.

☞ READ, AND JUDGE FOR YOURSELVES
On which side the grossest perjury and lying was perpetrated
ON THE TRIAL AT UTICA

The Phrenological Character of Mc'Leod was written as on page 5, in Utica before the close of the trial. Subsequent events are proving, and will yet prove, the correctness of the opinion there formed by F. Coombs. (Compare his extraordinary measures, with above one hundred others, in Coombs' Phrenology.)

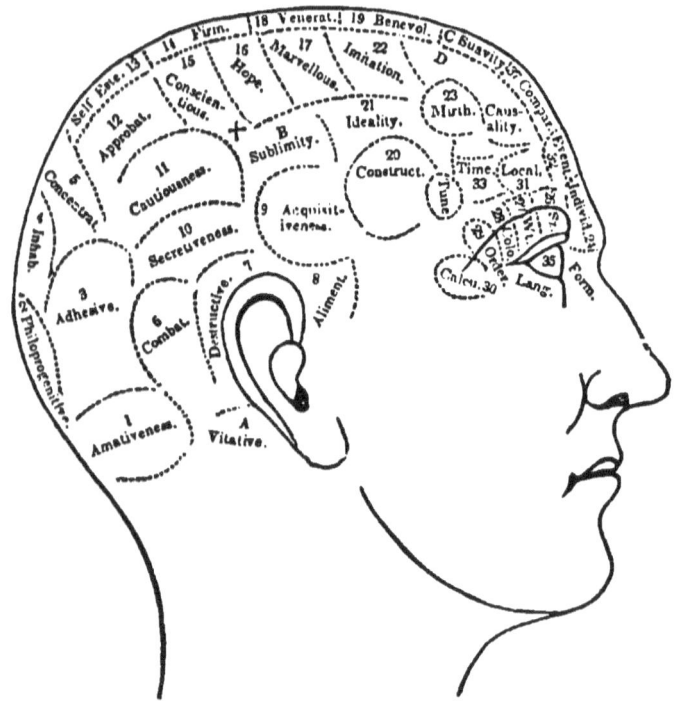

PHRENOLOGICAL HEAD,

EXHIBITING THE LOCATIONS OF THE ORGANS

A CORRECT PORTRAIT OF A. McLEOD.

MR. MC'LEOD'S OWN DESCRIPTION OF HIS MORAL CHARACTER,

Communicated to Dr. J. A. H. of New-York.

"I admit I am not what the world calls a *moral man.* Not that I am dishonest. No. When I was serving as sergeant in the 12th Lancers, at London, I got into debt, and lived on sixpence a day till I got out of debt. But I have kissed men's wives and daughters. I wont say so much about the wives, but I am pretty certain about the daughters. When Mrs. Taylor claimed my protection, I gave it, and got her lodgings. Then I thought that, as I had gone so far there could not be great wrong in my taking a *quid pro quo.* So you see that is the way this affair terminated."

PHRENOLOGICAL DESCRIP ION OF THE CHARAC
TER OF A RUFFIAN, PERHAPS A MURDERER—
BY FRED. COOMBS. PHRENOLOGIST.

Utica, Oct. 11, 1841.

The head of this person is larger in the base of the brain or the animal propensities, than any head the au.. or has measured or examined for several years, whilst the forehead or intellectual organs as well as the region of the moral sentiments, including both the directing and controlling organs, are very extremely deficient. He is so small in caution that he is scarcely susceptible of fear, and often rash, imprudent and very indiscreet, and does not hesitate to say or do any thing whatever to effect his object, particularly with females. He has no scruples of honor or conscience, and only thinks of his own gratification, regardless of every consideration of decency or honesty. He would openly boast of his immorality, and would make no pretensions to either virtue or religion. He would be an invaluable tool for loyal tories, or an excellent brigand. He is almost insensible to the softer emotions of either love or charity for his fellow men, and was scarcely ever known to perform a disinterested action. He is a perfect slave to his animal propensities, both for women and wine, or war ; and we have no hesitation in asserting, that such a head would not hesitate to kidnap, murder, or destroy his fellow beings, should there be a moderate inducement to do so. He could meet death in battle when his passions were roused, without fear, but not having much firmness or moral courage, he can conquer others much easier than himself. His animal courage is very great, and he is also fond of boasting of it or any other of his physical qualifications, and like George the fourth, he is a six bottle man ; and like him, would also spend more than his income, and consider the people legitimate prey. His deadliest enemies would be his own passions, and very great allowance ought to be made for a head of such an unfortunate organization. He is vicious, vindictive, jealous and revengeful, and ought always to be avoided The neighborhood of such a man would be a public misfortune, unless amongst lawless freebooters, with such he would be perfectly lionised. He would consider it a great misfortune to be absent from any marauding expedition, and if the plunder or expected emolument was considerable, it would be very difficult, or next to impossible, to keep him away He s an excellent monarchist if not a moralist, as he prefers living by plunder rather than honest labor, and would be a regular office hunter for spoils. He would be also fond of gunning or mimic warfare, and indeed of every exciting amusement, such as gambling, horse-racing, man-fighting, dog-fighting, &c.

It would be next to a miracle to find such a head conducting itself with either decency, honesty, or propriety, as it is strikingly and peculiarly like that of a murderer, whose measurement is given in Coomb's popular Phrenology, page 14, and history, page 113, and should this man be executed, he will also, in his last moments, conduct himself with perfect coolness and self-possession. He is a man of prodigious strength and courage; he is also very active and enduring in his physical organization, but not capable of much mental application, nor of reasoning clearly or philosophically on any subject. He has a keen eye to his own interest and makes money, but cannot keep it, as he is sure to indulge himself in expensive amusements so long as he has the means or credit to do so. He is so selfish and small conscientiousness, that he thinks every thing to be his own which he can lay his hands on.

This person ought to be taken great care of or he will do mischief, and as he is so very deficient of the moral or restraining faculties, he ought to be directed and restrained by others, and in order to gratify his natural propensities he ought to be a butcher or slaughterer of animals, (and not men) as in that capacity, he could gratify his feelings and be a benefit to the community. Such a man has almost a monomania to destroy or kill, and where he has exhibited this so strongly he ought not to be let loose on society, but to be taken great care of and have a keeper. Thus he might make amends for the wrongs and outrages he has committed on society; whilst to hang such a man is directly to impugn the divine author of his existence, who created and endowed him as he is. Delighting in war as much as a bull dog or tiger, he can never be an intellectually great, or a really sentimental good man; it is absurd to expect it from such an head, and the best we can do is to prevent them doing mischief. He is radically, and naturally, a bad man, and ought never to be trusted. A head of this description certainly deserves our pity as much as our hate, and to expect any high disinterested virtues from such a head would, indeed, be to expect figs from thorns or grapes from thistles. This head may be considered as one of that character which is naturally prone to evil or cruelty.—His low cunning is very great, as his secretiveness is large and having small conscientiousness, he would as soon tell a lie as the truth and much rather, if he is to gain any thing by it. This head is strikingly like that of Deaf Burke, the celebrated pugilist. He is also extremely similar to that individual whose person and features he bears a strong resemblance, being rather more ugly than the Deaf one. His physiognomical expression is extremely repulsive. He has a singularly large, strong and protruding under jaw. He has also, a particularly large mouth, and possesses a peculiarity in having the corners of it drawn down in a very remarkable degree, which occasionally gives him a very grim appearance. This peculiarity the writer's attention was confirmed in, by the remark of a stranger to a by-stander in the court, who observed that he had a perfect tiger like expression, and to communicate his idea applied his fingers and drew down each corner of his mouth. His face is immensely broad, but its disproportionate length is from below the nose, which feature is very short, and appears turned up, from having as it were, an immense broad lump on the end of it. This feature is extremely deficient in finish or delicacy of outline, and projects but a very short distance indeed from his face, whilst it is very broad at the base, having a bumpish or flat appearance. He is particularly broad across the brows, and he must be a first rate mathematician and ready calculator, and is, naturally, a good artist and mechanic, so far as exercised, and has an extraordinary good memory of persons, places and positions. Viewing him in front, the height and breadth of face is immense, whilst the forehead is very deficient indeed, in heighth, and being singularly broad does not improve his appearance. His head being composed of two thirds face, the greater portion of which is mouth, to signify he came to bite the world. His physical activity, strength and endurance must be very great, and he must be very fond of real or mimic warfare, as wrestling, boxing, shooting, or any athletic games, in which he would beat most competitors.

He would be a remarkably active partisan, and a very powerful one, on whichever side he could get the most pay from, as *might* with him constitutes *right*, he would be sure to hold with the strongest and richest party, and not being very discriminating, sentimental, or sensitive, it is easy to persuade such a man his interest is his duty. He would go any lengths of daring or desperation, in proportion to the reward held out to him, as he is almost insensible to fear in any case. His selfish propensities being enormous, he would generally be blind to all other considerations but selfish advancement. Such a man would be invaluable to the British tories, to carry out their schemes of wholesale plunder and murder; also, if necessary, he could be recommended for the discharge of any duty, however diabolical or hazardous, should there be no possibility of the adverse party giving him a greater sum to turn round and murder on the other side.

Phrenological Opinion on the talents and character of 1~r1 *Loyd Mackenzie, from actual measurement and examination, by*
FREDERICK COOMBS, PRACTICAL PHRENOLOGIST.

The head of this gentleman is very large indeed, as shown by measurement. He is rather a diminutive person, and does not weigh, probably, 120 pounds, whilst A. McLeod must be, nearly or quite, 200 pounds, which gives an immense difference in favor of the moral and intellectual character of Mackenzie.

A Phrenologist would infer from the head of Mackenzie, that he possesses intellectual powers of a very high order, but is deficient in the perceptive or observing faculties. He can plan far better than he can execute, and is also moderate in animal courage and nevei contends unless he is sustained by his moral sentiments and intellect, and can wield the pen better than the sword. He is singularly benevoient, kind, and charitable, and would be apt to spend money, time, and talents, for others, which he ought not to do if he regards his own interest, but this he too often forgets. He is remarkably enthusiastic, has very large hope and mirthfulness, and no troubles or difficulties annoy him long, as he has a remarkable flow of good spirits, and, troubles which would drive some men to desperation, are only subjects of ridicule for him.

His temperament or constitutional activity is of the very highest order, he can never be indolent or inactive. He is remarkably excitable and quick about every thing he says or does, and rather impatient but seldom revengeful or cruel, even when he has the power to be so, and very forgiving. The distresses of others annoy him as much as his own. He can never tolerate injustice, and would be ready to sacrifice every thing he has in the world to resist tyranny and unlawful aggression, whether offered to himself or others. He is too fond of his children and family, and would be rather apt to spoil them by indulgence, and would never be happy when absent from home and family. He is a most devoted friend and lover, extremely fond of the society of the fair sex, and is very gallant and polite, and would never be perfectly happy in the single state. He is a great favorite, and is very partial to the society of ladies; as he is remarkably social, and fond of refined and intellectual conversation.

His order and method are very good; and he has a time, mode, and place for every thing. He is ready in figures and mathematics.

His memory is very good, of generalities; but somewhat defectve, of names and details. He has good taste; is fond of the beautiful, the ideal, and believes in perfection; and, being apt to judge others by himself, very often gets deceived. He has a better opinion of mankind than they deserve. His innate feeling of justice is very strong; and he would be truly unhappy could he not discharge his obligations of a pecuniary nature. He is extremely sensitive to kindness, and would more than repay a favor conferred on him. His generosity is unbounded, and would generally keep him poor and penniless; he would be improvident in money matters.

He has a fair share of ingenuity in mechanics, and can generally make any thing he attempts, although he is more fond of literary than of mechanical pursuits. He is generally employed in writing, speaking, or thinking; and he generally reasons from first principles, and arrives at correct facts; but he is not so skilful in remedying as in discovering and denouncing abuses, of whatever nature, form, or quality.

He is petulant, and may domineer occasionally, but never do any thing knowingly wrong. He is a true patriot and philanthropist, and loves all mankind, exceping those whom he believes to be wicked, lying tories. He is a sincere republican, and will always judge the failings of others leniently; and if he is ever severe, it is that he may be just. Towards the majority or the poorer classes, and the unfortunate, he has too much sympathy, as he goes too great a length to serve them; particularly as he has been through much suffering himself, and can the more correctly appreciate the misfortunes of others.

He has not much brute courage, and not a very great degree of moral, or firmness in excess. His moral character is invariably good, excepting he might have some errors of youth yet remaining, but on the whole is rather exemplary, although not so religious as he might be. In this respect he is very accommodating, will always reason amicably, and is not sectarian, but very liberal, and never judges men by their professions, but by their actions. He has strong partialities and aversions, and is not always impartial in his judgement, as his feelings are very sensitive. He desires the good of all mankind, and never hesitates to avow his opinions, regardless of selfish considerations, and is occasionally indiscreet in this respect. He will make any sacrifice for the good of his family, and never leaves any stone unturned by which they may be benefitted. He is remarkably kind and affectionate in every relation of life, and will lend or give away almost any thing he has in the world. His reasoning faculties are remarkably acute and searching, as he always goes back to first principles, and reasons from cause to effect.

His conscientiousness is very large and he could not deliberately do wrong to another.

A. Mc'Leod now having left this country, the author conceives he may state facts respecting him, which came to his knowledge a the moment of his leaving Utica. The U——— S——— informed him that Mc'Leod had left Utica considerably in his debt; a great portion of which was money advanced for necessaries during his confinement; although, to show he was lately supplied with abundance of money from some quarter, he had champaign suppers very frequently at Bagg's Hotel, to say nothing of fees to counsel, &c. And yet this man left without even returning thanks to one who had so much befriended him in his necessity; in fact, it is perfectly notorious Mc'Leod was treated here more like a gentleman on a friendly visit, than a felon accused of murder.

The author might also state, as illustrative of his preceding remarks, that he obtained the measure of Mr. Mc'Leod's head three days before the close of the trial; that of Mr. Mackenzie, about the same time; and he was only prevented from giving publicity to them in their present form from the fear of prejudicing the cause of the then prisoner. Mr. Mc'Leod, at that time, and every day subsequently, promised the writer that he would sit to have his bust or cast taken; as he informed him, after measuring it, that his head was a very extraordinary one, and ought, by all means, to be preserved to the admirers of phrenology. And even so late as this morning, at half past eight o'clock, he promised to sit at ten or eleven o'clock, when, at the same time, his trunk was packed, and he started off by railroad, within half an hour of so promising. Thus, to the last moment, he was fully determined to earn the character previously written of him.

We must conclude by quoting the words of one of his own virtuous witnesses, after the trial, who observed, in our hearing, that, considering the very active part Mc'Leod had invariably taken immediately before and after the Caroline affair, it may be considered a miracle that he was not there. So miracles have not yet ceased, gentle reader.

As these very loyal captains, colonels, &c., pensioned on full pay by the British tories, are taught to believe, as part of their creed, their king or queen can do no wrong, and who have proved themselves their ready tools to commit murder on peaceable, unarmed American citizens, at midnight, it requires no great stretch of imagination to conceive they might so far obey their Sovereign's mandates, as to think it their duty to clear the prisoner by any means, foul or fair. See the trial

INCIDENTAL REMARKS.

We would humbly suggest to the American people, that the first maxim of Tory British Warfare is invariably to carry on the war in the enemy's territory. This, we believe, they have always done to the terror and cost of all nations; and we would now say, that the Americans have a most excellent opportunity of fairly retaliating, by sympathizing with all of these outraged and injured subjects of Great Britain, and their name is legion, as Chartists, &c. &c. Not that we would arm them with leaden bullets to murder their tyrants, but with paper bullets of the brain, by which they might estimate the difference of the governments, Republican and Monarchial, in the shape of cheap pamphlets, lectures, &c. And could something be contributed for this purpose by the American people, with funds or contribution, we have no hesitation in saying it would prove a greater safeguard to our shores, and to the liberty and independence of the country, than a whole navy of steamers, and at a hundredth, nay, a thousandth part of the cost; and without resorting to brute force and wholesale murder, which war necessarily must be, whether undertaken by single duelists, or by hosts of them, as we humbly conceive, there is just as much justice and legality in two men killing each other, as two nations; and war, either wholesale or retail, never yet facilitated or promoted justice; but, on the contrary, *might* invariably usurps the place of *right*. And hence the horrible cruelties and wars of the British tories. In every quarter of the world, they commence in fraud and terminate in murder. See the histories of America, India, China, &c. Also instanced in the case of the Caroline; its outrageous violation of our territory; the wanton and cruel murders of unarmed and defenceless men, at midnight; the destruction of the boat, &c., planned and executed by lawless desperadoes and hireling tories; and this act, as will be seen, to its chief actors is the high-road to nobility, (save the mark!) preferment, and favor, by these venal and corrupt tories; thus adding insult to injury on the American people. And let it be remembered, these tories at home are the origin and source of every villainy in Canada, &c., by their standing army of tory pensioners, governors, &c.; whose only mission or business in Canada is to plunder the people, under the name of governors, colonels, captains, &c.

Could Mackenzie, and other patriots, be liberally furnished with funds to circulate their papers in Canada, it would go further to destroy this tory power of injuring America than the far more costly mode of building steamers, erecting batteries, &c. Not but these are absolutely necessary in the present state of affairs ; but the poor of Great Britain, Canada, Ireland, &c., would be the best defenders of the American people, would the latter but take the pains to sympathize with them, and contribute small sums, either to Chartists, Repealers, or Patriots ; for all these are engaged in the same holy cause, LIBERTY for the people, and in destroying tory usurpation and robberies, which, under various names, as taxes, &c. &c., has been computed to exceed one million dollars per day, from the labors of the working classes in England, Ireland, and Scotland, alone. Hence the starvation, appalling crimes, and miseries there, to make us weep.

Americans ! have you any conscientiousness, or can you, by your silence, approve of such atrocities?

It is almost needless here to repeat the fact, that their poor people of England have no more to do with the acts of the government, or are responsible for its outrageous tyrannies and robberies, than are the American people themselves. This is a fact which ought for ever to be kept in view by the American people. And also, that the tory government of England are almost as much hated and feared by the great bulk of the people in England, as they are by other victims—the Irish, the Chinese, East Indians, &c. ; and, indeed, they have much greater cause to fear and hate them, for they are robbed to an infinitely greater extent than all other foreign nations whatever, put together. The Canadians themselves are not robbed probably above one tenth as much as the poor English, Irish, &c., at home.

The gallant Chartist, O'Connor, just released from prison, says that he shall have four millions of Chartists' signatures within six months. These, added to O'Connell's Repealers, the Corn Law Repealers, and Patriots in Canada, will give these tories full employment at home ; and we think the Americans have not much to fear from them. These agitators are our present best defenders. Let us help them.

What we desire above all things is to impress on the American people the natural sympathy the people of England, Ireland, &c., have for them and republicanism, and hatred of their own thievish, tyrannical government ; and with a little assistance, or even without it, we are confident the English and Irish will be enabled, sooner or later, to free themselves from their present odious slavery. Americans would be immediately benfitted by their so obtaining it.

MURDEROUS ATTACK ON THE AMERICANS.

FROM THE FREEMAN'S CHRONICLE, PAGE 107.

The steamboat Caroline took out a license at Buffalo, as a ferry boat for passengers—sailed to Tonawanda—thence to Schlosser, and twice between it and Navy Island. Schlosser contains an old store-house and a small inn. At five o'clock in the evening the Caroline was moored at the wharf. The tavern being very full, a number of gentlemen took beds in the boat—in all about thirty-three persons slept there. A watch was placed on deck at eight o'clock; the watchmen unarmed—there was only one pocket pistol on board, and no powder. At midnight, the Caroline was attacked by five boats, full of armed men from the English army at Chippewa, who killed (as themselves say) six men, or as the American account has it, eleven. A number was severely wounded, as the people in the American port could make no resistance. To kill them was, therefore, a wanton assassination. The cry of the assailants was " G—d d—n them—no quarter—fire, fire!"

Amos Durfee, of Buffalo, was found dead upon the dock, a musket ball having passed through his head. The Caroline sailed under the American flag, which the assailants took to Toronto, and displayed at annual festivals, in honor of this outrage. She was set in a blaze, cut adrift, and sent over the falls of Niagara. We witnessed the dreadful scene from Navy Island. The thrilling cry ran around that there were living souls on board; and as the vessel, wrapt in vivid flame, which disclosed her doom as it shone brightly on the water, was hurrying down the resistless rapids to the tremendous Cataract, the thunder of which, more awfully distinct in the midnight stillness, horrified every mind with the presence of their inevitable fate; numbers caught, in fancy, the wails of dying wretches, hopelessly perishing by the double horrors of a fate which nothing could avert; and watched, with agonized attention, the flaming mass, till it was hurried over the falls to be crushed in everlasting darkness in the unfathomed tomb of waters below. Several Canadians, who left the Island in the Caroline that evening, to return next day, have not since been heard of, and doubtless were among the murdered, or hid on board and perished with the ill-fated vessel.

AN APPEAL FOR THE MEXICANS.

The author of the preceding pages having, several years previous to this expression of his sentiments, renounced allegiance to the British Government, and having since that period revisited his early home and several British Colonies, he still repeats, and has been from actual observation fully satisfied of the monstrous and cruel injustice of that Government to all classes of its subjects, excepting, perhaps, the very few recipients of its favors, and the fewer privileged aristocratic classes, who are already encumbered with enormous and prodigal wealth, and for the most part used only for the worst of purposes, and requiring no prophet to predict the speedy downfall of such an accursed system.

Holding up to the view of the world, the sublimely moral spectacle, that neither nations or people can outrage nature's laws with impunity, and one of the greatest and most flagrant crimes the English Government has perpetrated since the Norman Conquest, (alias Butchery,) has been in butchering other nations and people, with whom they had either real or imaginary difficulties.

They have been consequently most eminently a fighting people; and in the language of one of her noble poets, speaking of one of her warriors, he says:

"He was the son of that true mother,
Who butchered half the world, and bullied t'other."
ASIA AND AMERICA.

This naturally leads every lover of republicanism and the well-being of America, by every means possible to avoid the disastrous and truly lamentable condition in which we find her placed; and as like effects follow like causes, are not the American people sowing tares and thistles, which they or their children will eventually reap in their present unhappy contest with a sister republic?

The writer, for one, greatly fears it, indeed, and thus entreats, aye, humbly entreats, the American Sovereigns, the mistaken people, to pause in their career of conquest, and view the fearful picture of impoverished and bankrupt Europe, who have squandered mountains of treasure, and shed oceans of blood in accursed war, and while huzzaing for conquest have undermined their own otherwise happy homes, in which they will, we fear, be finally engulphed.

Well, beloved Americans, let us implore you, in the name of all that is great and good, to abjure this war; aye, at all hazards, to stop this war; yes, we most humbly implore you as the veriest beggar, we beg and entreat you to act with mercy and forbearance to a conquered and greatly mistaken, misruled people, the Mexicans. Act with that imperishable and glorious magnanimity, and noble generosity, you exhibited towards our common brethren, the Irish and Scotch. So shall your name ascend as grateful incense before high heaven.

"The quality of mercy is not strain'd;
It droppeth, as the gentle dew from heaven,
Upon the place beneath. It is twice bless'd;
It blesseth him that gives, and him receiving:
'Tis mightiest in the mighty; it becomes
The throned monarch, better than his crown,
It is an attribute to God himself,
And earthly power doth then show likest **God's**,
When mercy seasons justice."

New York, Nov. 1, 1847.

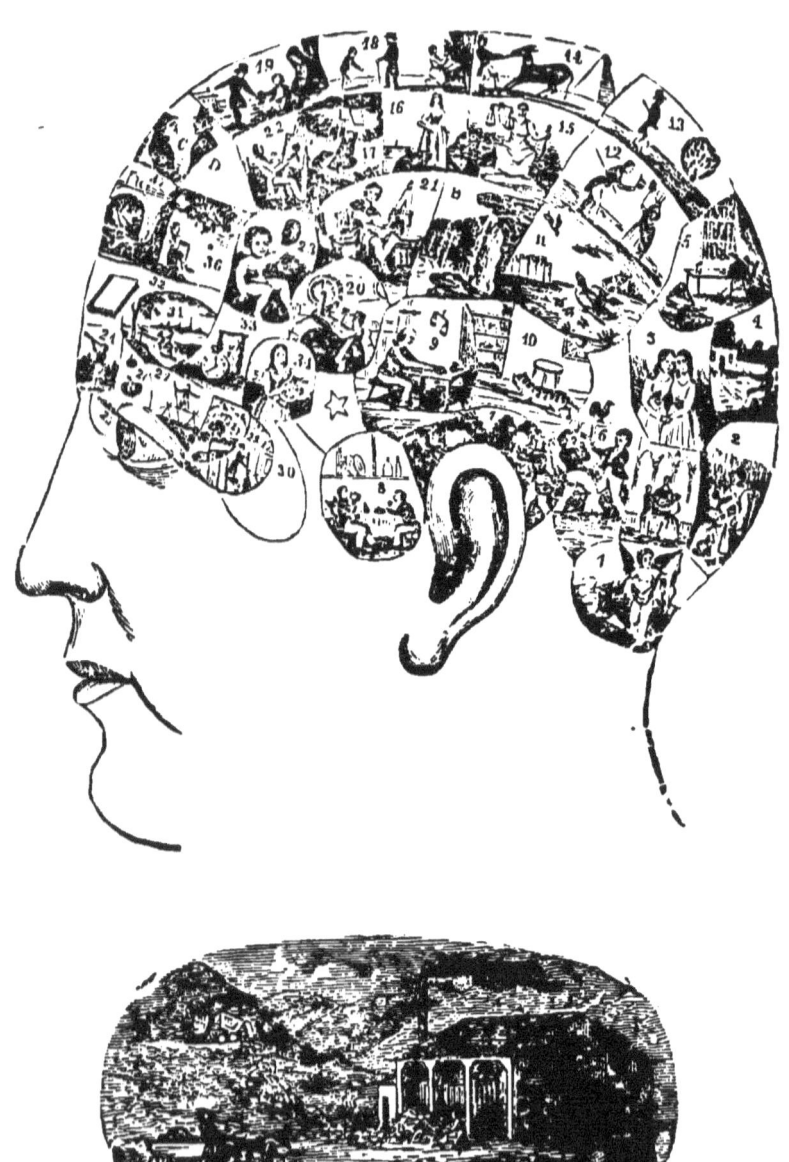

www.ingramcontent.com/pod-product-compliance
Lightning Source LLC
Chambersburg PA
CBHW020127170426
43199CB00009B/675